Towards a Unified Soil Mechanics Theory

Authored by

Eduardo Rojas
Universidad Autónoma de Querétaro, México

Towards a Unified Soil Mechanics Theory:

The Use of Effective Stresses in Unsaturated Soil, Revised Edition

Author: Eduardo Rojas

ISBN (Online): 978-1-68108-699-6

ISBN (Print): 978-1-68108-700-9

© 2018, Bentham eBooks imprint.

Published by Bentham Science Publishers – Sharjah, UAE. All Rights Reserved.

General:

1. Any dispute or claim arising out of or in connection with this License Agreement or the Work (including non-contractual disputes or claims) will be governed by and construed in accordance with the laws of the U.A.E. as applied in the Emirate of Dubai. Each party agrees that the courts of the Emirate of Dubai shall have exclusive jurisdiction to settle any dispute or claim arising out of or in connection with this License Agreement or the Work (including non-contractual disputes or claims).
2. Your rights under this License Agreement will automatically terminate without notice and without the need for a court order if at any point you breach any terms of this License Agreement. In no event will any delay or failure by Bentham Science Publishers in enforcing your compliance with this License Agreement constitute a waiver of any of its rights.
3. You acknowledge that you have read this License Agreement, and agree to be bound by its terms and conditions. To the extent that any other terms and conditions presented on any website of Bentham Science Publishers conflict with, or are inconsistent with, the terms and conditions set out in this License Agreement, you acknowledge that the terms and conditions set out in this License Agreement shall prevail.

Bentham Science Publishers Ltd.
Executive Suite Y - 2
PO Box 7917, Saif Zone
Sharjah, U.A.E.
Email: subscriptions@benthamscience.org

CONTENTS

FOREWORD

The choice of stress variables controlling the behavior of unsaturated soils has been a challenge for geotechnical engineers for more than half a century and in particular in the last two decades. Many researchers in the world are still working on this aspect and Professor Eduardo Rojas is one of them. There are two main approaches: the first one considers two stress variables, generally the net stress ($\sigma-u_a$) and matric suction (u_a-u_w); the other considers a single stress variable, generally called the effective stress. However, this latter approach generally shows deficiencies in reproducing the phenomenon of collapse under wetting.

In several technical papers and in this eBook, Professor Eduardo Rojas has developed a porous-solid model that considers micropores, macropores and their connections, and the fact that they can be saturated, unsaturated or dry. This model is very powerful and allows the determination of the Soil-Water Retention Curve on both the drying and the wetting paths. This model is also the key for the determination of an equivalent effective stress that allows the analysis of the different aspects of the mechanical behavior of unsaturated soils: compression, strength, tensile strength and volumetric behavior. Application of the model to experimental results published in the literature gives remarkable results.

Was it necessary to put this information into one eBook? The answer is yes, as it shows in one document the continuity from the physical model and the definition of an equivalent effective stress to the practical applications. Did the eBook provide a final response to the questions concerning representative stress variables for unsaturated soils? Probably not, but it gives a very consistent approach of the problem within the context of the existing literature and knowledge. It is a reference for all those interested in the behavior of unsaturated soils.

Serge Leroueil
Laval University
Quebec, Canada

PREFACE

With the introduction of the effective stress concept, the behavior of saturated soils could be clearly understood and the basic principles of the saturated soil mechanics could be established. The effective stress principle states that the strength and volumetric behavior of saturated materials are exclusively controlled by effective stresses. Constitutive models for saturated materials of different types are all based on the effective stress principle. Later, the Critical State theory combined the strength and volumetric behavior of saturated soils in a simple and powerful constitutive model. A great number of models for saturated soils are based on the critical state theory.

Things went not so smooth for unsaturated materials. More than fifty years ago Alan W. Bishop proposed an equation for the effective stresses of unsaturated soils. However, this equation was severely criticized because it could not explain by itself the phenomenon of collapse upon wetting of soils. In addition, Bishop's effective stress parameter χ showed to be extremely elusive and difficult to be determined experimentally. Given these difficulties the use of the so-called independent stress variables (mainly net stress and suction) became common in unsaturated soil mechanics. Different equations for the strength and volumetric behavior of soils were proposed and the theory for unsaturated soils became distant from that of saturated materials. The Barcelona Basic Model represents one of the most simple and completes models within this trend. The Barcelona Basic Model enhanced the Critical State theory to include unsaturated materials and give a plausible explanation to the phenomenon of collapse upon wetting. This model proved that this phenomenon could only be modeled if in addition to a volumetric equation, a proper elastoplastic framework was included. But then, the simulation of some particular phenomena related to the strength and volumetric behavior of unsaturated soils, appeals for the inclusion of the hysteresis of the soil-water retention curve and the hydro-mechanical coupling observed in unsaturated materials. The difficulties met in introducing these aspects into the independent stress variables models made it clear that a different approach should be considered. Then, quietly elastoplastic models based on Bishop's effective stress equation started to appear showing its superiority by including the hysteresis of the soil-water retention curve and the hydromechanical coupling of unsaturated soils. And finally, the debate about the appropriateness of Bishop's equation to represent the effective stresses for unsaturated soils is slowly coming to an end. This transformation on the construction of constitutive models for unsaturated soils is also leading towards a unified soil mechanics theory.

This book shows how the effective stress principle can be applied to simulate the strength and volumetric behavior of unsaturated soils employing the same equations commonly used for saturated materials. The book initiates with an analysis of the stresses transmitted to the different phases of an unsaturated soil when it is loaded. This analysis results in an expression for the stresses carried by the solid skeleton of the material. This expression can be written in the same terms as Bishop's effective stress equation and leads to an analytical expression for Bishop's effective stress parameter χ. However, the variables required to obtain χ cannot be experimentally determined. For that purpose, a network porous-solid model is developed which is able to approximately reproduce the structure of soils based on the grain and pore size distributions of the material. This porous-solid model is able to determine the allocation of water in the pores of the soil and thus simulate the soil-water retention curves of the material including the scanning curves. It is also possible to obtain the required parameters to determine the value of Bishop's parameter χ and therefore compute the current effective stress. Nevertheless, the use of a network porous-solid model requires large memory storage capacity that cannot be presently found in common computers and thus the porous-solid model computer program becomes very slow. For that reason, a probabilistic porous-solid model is developed that reduces the storage requirements and speeds the simulations. Finally, an elastoplastic framework is developed to account for the volumetric behavior of unsaturated soils including compacted materials. This framework allows the simulation of the collapse upon wetting phenomenon and explains some other phenomena that could not be explained using the independent stress variables approach. All these developments lead to a general framework for the strength and volumetric behavior of soils including saturated, unsaturated and compacted materials. In that sense, a unified soil mechanics theory is presently on its way.

Eduardo Rojas
Universidad Autónoma de Querétaro
México

ACKNOWLEDGEMENTS

The construction of knowledge is a collaborative task; that is why I would like to acknowledge all those who have contributed in one way or the other with their analysis, opinions, experimental results, comments and criticisms to this book. Some of them are listed in the references although, this list is not exhaustive.

All graphical material contained in this book was carefully prepared by María de la Luz Pérez-Rea and Hiram Arroyo Contreras, Researcher and PhD student at the School of Engineering of the University of Querétaro, Mexico, respectively. Their help is greatly appreciated.

CONFLIC OF INTEREST

The author confirms that the eBook content has no conflict of interest.

Eduardo Rojas
Universidad Autónoma de Querétaro
México
E-mail: erg@uaq.mx

DEDICATION

To Silvia, my loving wife, and my son and daughter:
Carlos Javier and Sonia Itzel, who have showed me the pleasure of shearing and enjoy
each other.

SUMMARY

When Karl von Terzaghi applied the effective stress principle in the soil mechanics theory, the strength and volumetric behavior of saturated soils could be clearly understood and general constitutive models for these materials could be developed. This book shows that the principle of effective stress can also be applied to unsaturated soils and that the same equations used to determine the strength and volumetric behavior of saturated soils can be applied to unsaturated materials. These developments open the door for general constitutive models that include saturated, unsaturated and even compacted materials leading to a unified soil mechanics theory.

CHAPTER 1

Introduction

Abstract: The use of the effective stress principle led to a general theory for the strength and volumetric behavior of saturated soils. Presently, all constitutive models for saturated soils are based on this principle. In 1959, Bishop proposed an equation for the effective stress of unsaturated soils. However, it was severely criticized because it could not explain by itself the phenomenon of collapse upon wetting. Moreover, an analytical expression for the determination of its main parameter χ was not provided and in addition, its value could not be easily determined in the laboratory. Since then several equations to determine the value of parameter χ have been proposed. Fifty years later, it has been acknowledged that Bishop's effective stress equation can be employed to simulate the behavior of unsaturated soils when it is complemented with a proper elastoplastic framework.

Keywords: Air pressure, Collapse, Constitutive model, Effective stress, Effective stress parameter, Elastoplastic framework, Independent stress variables, Pore water pressure, Saturated soil, Shear strength, State surface, Suction, Total stress, Unsaturated soil, Volumetric behavior.

1.1. DIFFERENT APPROACHES FOR UNSATURATED SOILS

Even though the idea of using effective stresses in the study of unsaturated materials is old, the incapacity to provide an explanation to the phenomenon of collapse upon wetting (among other reasons) made this approach to be abandoned for about forty years. During that time some other approaches to study the behavior of unsaturated soils were used. The state or constitutive surfaces [1], as the one represented in Fig. (**1**), were used for some time. In these plots, the behavior of a certain state variable, as for example the void ratio, is plotted as a function of two independent stress variables mainly the mean net stress $(\bar{p} = p - u_a)$ and suction$(s = u_a - u_w)$,where p represents the total mean stress and u_a and u_w the air and water pressures, respectively. This procedure aimed to establish mathematical relationships between the void ratio or the degree of saturation with the independent stress variables as Hung, Fredlund and Pereira [2] have done. This method represented to some researchers the acceptance of the inexistence of an effective stress equation for unsaturated materials (see for example [3]). However, state surfaces soon showed their limitations. For example,

Eduardo Rojas

unicity could only be ensured under certain conditions especially because of the hysteresis of the soil-water retention curve (SWRC), the hydro-mechanical coupling and the dependency of soil behavior on the stress-path. In any case, this task would have been formidable complex because the behavior of unsaturated materials depends not only on the mean net stress and suction but also on the degree of saturation and the structure of soils. Recently Zhang and Lytton [4] proposed a modified state-surface approach under isotropic stress conditions that can be applied to the study of the volumetric behavior of unsaturated soils including collapsing and expansive soils.

Sometime later the independent stress variables approach was employed to study the behavior of unsaturated soils. The independent stress state variables were defined as those stresses controlling the strength and volumetric behavior of soils. By performing the analysis of the equilibrium of an elemental volume of unsaturated soil, Fredlund and Morgenstern [5] proved that the use of two out of three possible combinations of the stress variables represented by the total stress (σ), the air pressure and the water pressure, were sufficient to completely define the state of stresses of an unsaturated sample. The three possible combinations are: $(\sigma\text{-}u_w)$ with $(u_a\text{-}u_w)$; $(\sigma\text{-}u_a)$ with $(\sigma\text{-}u_w)$; and $(\sigma\text{-}u_a)$ with $(u_a\text{-}u_w)$. Being this last combination, net stress $(\sigma = \sigma\text{-}u_a)$ and suction, the most employed to study the behavior of unsaturated soils.

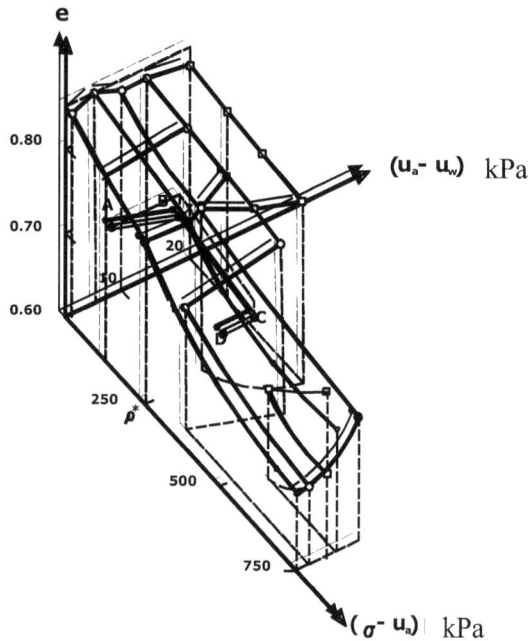

Fig. (1). State surface for the void ratio (after [1]).

This theoretical analysis co-validated the experimental observations made by Bishop and Donald [6] in 1961. These researchers performed a series of triaxial tests where the confining stress (σ_3), the air and the water pressures were all independently controlled during the loading of the sample. In this way the values of the net confining stress (σ_3-u_a) and suction could be maintained constant throughout the test while the independent pressures could change. These results showed that the independent variations of σ_3, u_a and u_w had no effect on the stress-strain curve whenever the confining net stress and suction remained constant. However, a variation on these values resulted in marked changes in the stress-strain curve of the sample.

With the use of the independent stress variables, the representation of the failure surface for unsaturated soils required, additional to the normal net stress (σ_n-u_a) and the shear stress (τ) axes, the inclusion of the suction axis as indicated in Fig. (2). This figure shows the failure lines for a saturated material (indicated by the friction angle φ) and for an unsaturated one (indicated by the friction angle φ_s) where for the last, the cohesion (c) appears as a strength parameter.

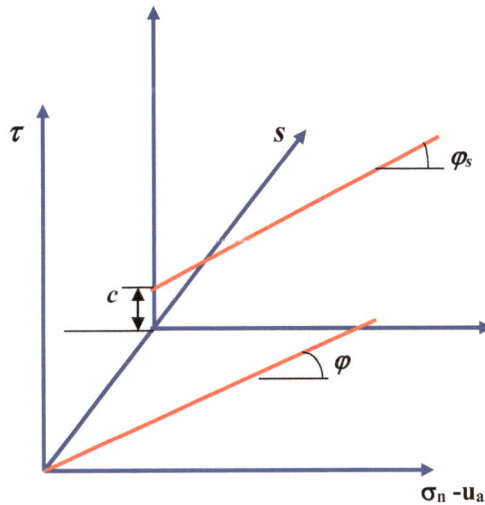

Fig. (2). Failure lines for the saturated and unsaturated conditions.

Following this tendency, Alonso, Gens and Josa [7] developed a constitutive model for unsaturated soils based on the modified Cam-Clay model (MCCM) developed by Roscoe and Burland [8]. This model, known as the Barcelona Basic Model (BBM), is one of the most simple and complete models to simulate the behavior of unsaturated soils including collapsing and expansive soils. One of the main contributions attributed to the BBM is that it clearly explains the phenomenon of collapse upon wetting by introducing the loading collapse yield surface (LCYS) as illustrated in Fig. (**3**). This phenomenon occurs when a saturated sample is dried (path AB in Fig. **3**), then loaded by increasing the net stress (path BC) and finally wetted up to saturation (path CD).

This behavior can be explained in the following terms: when a soil sample dries, it stiffens because additional contact stresses between solid particles appear due to the development of water menisci. Then the apparent preconsolidation stress increases and therefore, the soil behaves as a preconsolidated material. Therefore, when the sample is loaded by increasing the net stress, it slightly deforms. Subsequently, when the soil wets, the additional contact stresses between solid particles reduce along with the apparent preconsolidation stress. Then, the volumetric deformation that did not occur during the loading stage when the soil was dry, takes place suddenly during the wetting of the sample. This means that when the sample is fully saturated, it returns to the saturated compression line. The inability of Bishop's equation to explain this behavior by itself was one of the major reasons to abandon this equation during several decades.

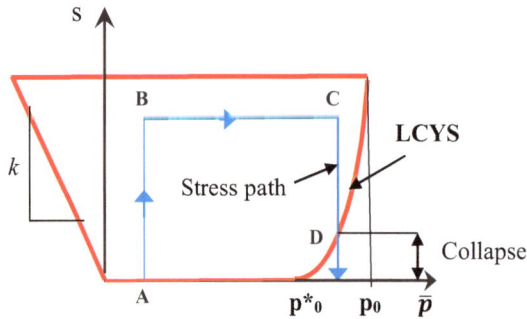

Fig. (3). Simulation of the phenomenon of collapse upon wetting by the inclusion of the LCYS (after [7]).

1.2. EFFECTIVE STRESSES

In 1936, Terzaghi [9] stated the principle of effective stress for saturated soils leading to the equation

$$\sigma' = \sigma - u_w$$

where σ' represents the effective stress. This equation implicitly considers the following two hypotheses:

a. Solid particles and water are incompressible.
b. The contact area between two particles is independent of the confining pressure and can be neglected.

If one of these hypotheses is missing, then different equations can be obtained. For example, if the contact area between particles is considered, the stress regulating the shear strength of soils [10] can be written as

$$\sigma' = \sigma - k u_w$$

where

$$k = \left(1 - a \tan \psi / \tan \varphi\right)$$

a represents the contact area between particles per unit area, ψ is the friction angle of the mineral comprising the solid particles and φ is the internal friction angle of the granular media.

On the other hand, according to Lade and De Boer [11], if the compressibility of the solid particles is considered, the value of parameter k for the volumetric behavior of saturated porous media is

$$k = \left(1 - (1 - n) C_s / C_e\right)$$

where n represents the soil porosity, C_s is the compressibility of the solid material comprising the solid particles and C_e is the compressibility of the soil structure.

The above expressions show that an effective stress does not represent a physical measurable quantity but it is an artificial stress used to simplify the relations for volumetric and strength behavior of materials and may include mechanical properties or state variables. In other words, it represents a constitutive variable. However, for the range of stresses frequently used in geotechnical engineering, the variation of parameter k is so small that it is very difficult to determine, even with sophisticated laboratory equipment. Therefore, it can be said that Terzaghi's effective stress equation represents an excellent approximation for both the shear strength and the volumetric behavior of saturated soils.

Because of this simplification, when researchers were looking for an effective stress equation for unsaturated soils, it was assumed that such equation should be written as a function of stress variables only and this assumption gave rise to a great deal of confusion.

In the late 50's some researchers focused on the behavior of unsaturated soils and proposed different equations for the effective stress: Jennings [12], Croney, Coleman and Black [13], Bishop [14] and Aitchison [15] among others. However, only that proposed by Bishop [14] prevailed. This equation writes,

$$\sigma' = \sigma - u_a + \chi\left(u_a - u_w\right) = \overline{\sigma} + \chi s \qquad \textbf{(1.1)}$$

where χ is a parameter mainly related to the degree of saturation (S_w).

Different expressions have been proposed for the value of parameter χ. For example Aitchinson [15] proposed the following relationship

$$\chi = S_w + \frac{1}{s}\sum_{i=0}^{s} 0.3\, s_i\, \Delta S_{wi}$$

This equation considers that parameter χ is a function of the addition of the product of the increment of the degree of saturation (ΔS_{wi}) multiplied by the value of suction (s_i) along the SWRC from suction cero to the current suction of the soil. This means that parameter χ is not only related to the degree of saturation of the material but also to the way water intrudes the pores of soil. In other words, the structure of soil also plays a role in the value of χ. A similar conclusion was reached by Jennings and Burland in 1962 when they reported that the void ratio also affects the value of parameter χ.

Later Blight [16] proposed two experimental methods to determine the value of parameter χ: the first one is based on the comparison of results of two triaxial tests, one performed on a saturated and the other on an unsaturated sample. The second one, results from the analysis of contact forces between two solid particles linked by a meniscus of water. However, the author where unable to conclude which method was the most suitable. Chapter 2 shows that this last method gives light on the value of parameter χ.

Recently, Khalili and Khabbaz [17] proposed the following equation to determine the value of parameter χ

$$\chi = \left[\frac{s}{s_{ae}}\right]^{-0.55}$$

where s_{ae} represents the air entry value. For suctions below the air entry value, it is considered that air is only present in the form of air bubbles, therefore $u_a = u_w$ and, Bishop's equation reduces to Terzaghi's effective stress equation. One important aspect of this equation is that it includes a parameter from the SWRC. The SWRC represents a relationship between the water content or the degree of saturation of the sample with suction. This trend, where parameters of the SWRC are used to obtain parameter χ, has been followed by other researchers with fair results as shown below.

Based on experimental evidence Öberg and Sällfors [18] proposed that, for granular materials and degrees of saturation over 50%, parameter χ may adopt the value of the degree of saturation (S_w). In this way, the simplified version of Bishop's equation appeared. Some researchers have proposed other empirical expressions for parameter χ based on the results of tests made on sand, silt and clay. Amongst the most successful are those shown in Table **1**. It is interesting to observe that all these expressions are closely related to the SWRC. These equations along with some others were confronted with the experimental results of different soils collected by Garven and Vanapalli [19]. The results of this exercise showed equation T1 as the most successful with 70% of success followed by equations T2 and T3 with only 25% and 17% of success, respectively. Even though equation T1 had a good rate of success, its major drawback is that it cannot account for the behavior of all types of soils as stated by Garven and Vanapalli [19].

Additional experimental results showed that the value of parameter χ was affected by different factors such as the wetting-drying history, the void ratio and the structure of the soil ([6, 20]).

Table 1. Some relationships for the value of parameter χ.

Number	Equation	Author
T1	$\chi = (S_w)^k$ k = fitting parameter	Vanapalli, Fredlund, Pufahl, and Clifton [21]
T2	$\chi = (S_w\text{-}S_r)/(1 - S_r)$ S_r = residual degree of saturation	Vanapalli, Fredlund, Pufahl, and Clifton [21]
T3	$\chi = S_w$	Öberg and Sällfors [18]

Added to the problem of the determination of parameter χ, the validity of Bishop's equation was questioned because it could not predict by itself the phenomenon of collapse upon wetting [20]. During this phenomenon, the volume of a soil sample suddenly reduces while the mean net stress remains constant. Therefore, intuitively, this phenomenon was interpreted as the result of an increment of the effective stress applied to the soil sample, while Bishop's equation predicts the reduction of this stress during wetting because suction decreases and becomes nil at saturation.

However, it is now known that, because collapse represents a plastic volumetric response of the soil, it can only be explained when an elastoplastic framework similar to that proposed by Alonso, Gens and Josa [7] is added to the constitutive model based on the independent state variables approach. This elastoplastic framework considers that the apparent preconsolidation stress of the soil reduces with suction. Therefore, collapse cannot be explained using a single constitutive variable as that represented by Bishop's effective stress equation without an elastoplastic framework.

Only recently Bishop´s equation has reappeared on the constitutive modeling for unsaturated soils as it has proven major efficiency in coupling the hydraulic and mechanical behavior of unsaturated materials (see for example [22 - 25]).

Although some attempts to obtain Bishop's effective stress equation have been done over the years (see for example [1, 18, 26]) none of them have prevailed. In the next chapter, a procedure to obtain an analytical expression for parameter χ is presented.

CHAPTER 2

The Effective Stress Equation

Abstract: Based on the analysis of the equilibrium of solid particles of an unsaturated sample subject to certain suction it is possible to establish an analytical expression for Bishop's parameter χ. The resulting stress can be used to predict the shear strength and volumetric behavior of unsaturated soils. The effective stress is written as a function of the net stress and suction and requires three parameters: the saturated fraction, the unsaturated fraction and the degree of saturation of the unsaturated fraction of the sample. This equation clarifies some features of the strength of unsaturated soils that up to now had no apparent explanation. A drawback to this expression is that the determination of these three parameters cannot be made from current experimental procedures.

Keywords: Degree of saturation of the unsaturated fraction, Dry fraction, Effective stress parameter, Effective stress, Equilibrium, Homogeneous material, Macrostructure, Microstructure, Saturated fraction, Shear strength, Suction, Total stress, Unsaturated fraction, Volumetric behavior, Water menisci.

2.1. INTRODUCTION

Most natural soils show a bimodal structure consisting in a microstructure and a macrostructure [27]. The microstructure can be formed by packets of fine particles that flocculate and remain attached. These packets or aggregates contain the intra-aggregate pores which are pores of small size. On the other hand, the macrostructure is the arrangement of packets of fine particles alone or with solid grains that show the inter-aggregate and inter-particle (when solid grains are present) pores which are pores of larger size. In such a case, the size of pores usually ranges from 500μm to 0.01μm. The smallest pores being close to the thickness of the adsorbed water layer which means that these pores never dry. This phenomenon accounts for the difference in the consistency of fine and coarse materials when dry. When suction applied to the soil is low, great part of the macrostructure and the totality of the microstructure remain saturated. When suction increases, the saturated soil volume decreases in such a way that some solids are now completely surrounded by dry pores while others are only partially surrounded by saturated pores. Instead most of the microstructure is still saturated.

Eduardo Rojas

Finally, for very high suction, the saturated soil volume tends to disappear while the dry fraction increases. In the case of coarse materials the saturated fraction may completely disappear while for clayey soils this never happens because of the existence of intra-aggregated voids filled with layers of adsorbed water. Therefore it can be said that, in general, an unsaturated soil consists of a saturated fraction, where soil particles are completely surrounded by water, an unsaturated fraction, where solid particles are linked together by water menisci and a dry fraction where solids are completely surrounded by air. In some cases, the bimodal structure may not appear for example in homogeneous dense sands. In that case, the transit from the saturated to the dry condition occurs very fast and the saturated fraction completely disappears at small values of suction while the dry fraction increases rapidly. This behavior is reflected on the soil-water retention curves (SWRCs) of every material as will be shown later.

If a soil sample is confined in a closed environment at a constant temperature during an appropriate period of time, then it can be admitted that the relative humidity is the same everywhere in the sample and therefore, the value of suction is constant throughout the sample. Thus, air and water pressures in the saturated zones are the same as for the unsaturated. This implies that all saturated zones are surrounded by menisci of water showing the same radius of curvature as the unsaturated zones.

2.2. EFFECTIVE STRESS EQUATION

Consider a homogenous and isotropic soil showing a bimodal structure where pores are randomly distributed as shown in Fig. (1). The term homogenous means that a representative elementary volume can be used to model the whole material as this volume adequately reflects both the microstructure and macrostructure of the system. The term isotropic means that the mechanical and geometrical properties are the same in all three directions, including the spatial distribution of menisci.

The solid particles constituting both the macro and the microstructure can be observed in Fig. (1). Also, the water menisci and gas phase are included. In general, it is considered that the solid particles of the microstructure are grouped in the form of packets. In this case, the influence of the contractile skin is ignored as both Haines [28] and Murray [29] demonstrated that its influence could be ignored for practical purposes. Also, the water vapor, adsorbed water and dissolved air are disregarded as Murray [29] has proved that their influence is also minimal. Finally, the contact areas between solids will be neglected as implicitly considered in Terzaghi's effective stress equation. Based on a Disturbed State Model, Desai and Wang [30] performed an analysis of the effective stress on saturated soils which includes the effect of the variation of the contact area of solids. A similar procedure could be used herein if the contact area of solids was not neglected.

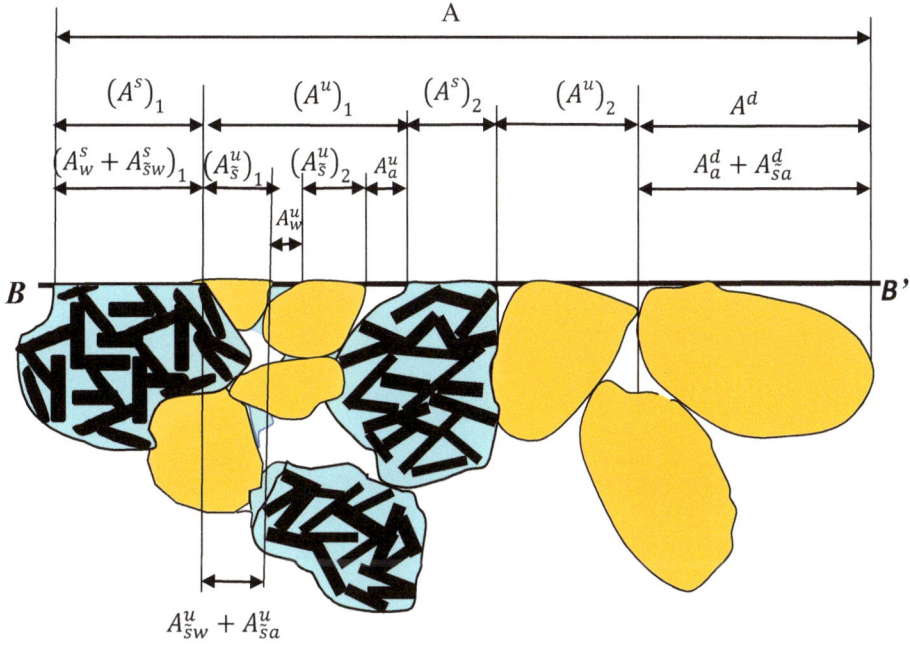

Fig. (1). Section of an unsaturated soil showing the contact areas of the different phases.

For this analysis, the following notation is used: a superindex indicates the fraction being referred: s for the saturated, u for the unsaturated and d for the dry fraction of the soil. A subindex indicates the phase being referred: $\hat{s}$ for solids, w for water and a for air. A double subindex indicates the influence of one phase to another; for example, $A_{\hat{s}a}$ and $A_{\hat{s}w}$ represents the area of solids subjected to air and water pressure, respectively.

Considering a unitary thickness of the soil section shown in Fig. (**1**), it can be established that the total area (A) of the cross section B-B', results from the addition of the saturated (A^s), the unsaturated (A^u) and the dry fractions (A^d) of the sample, that is to say $A = A^s + A^u + A^u$. Also, the total area of the saturated fraction results from the addition of the area where water directly reacts (A_w^s) plus that occupied by solids ($A_{\hat{s}}^s$), in the form $A^s = A_w^s + A_{\hat{s}}^s$. Moreover, the solid particles on the saturated fraction are only in contact with water and other solids. If the contact area between solids is neglected, then all the horizontal projection of the area of solids represented in section B-B' is subject to the water pressure; that is to say $A_{\bar{s}}^s = A_{\bar{s}w}^s$. Therefore, the total area of the saturated fraction can be written as the sum of the areas where water directly reacts and the horizontal projection of solids pushed by water; that is to say

$$A^s = A_w^s + A_{\bar{s}w}^s \tag{2.1}$$

On the other hand, the total area of the unsaturated fraction results from the sum of the areas where the solid ($A_{\bar{s}}^u$), liquid (A_w^u) and gas (A_a^u) phases react, that is to say $A^u = A_{\bar{s}}^u + A_w^u + A_a^u$. Additionally, the solids also are in contact with the three phases. If the contact area between solids is ignored, then the horizontal projection of the solids on section B-B' results from the addition of the areas of solids where the pressures of liquid ($A_{\bar{s}w}^u = \left(A_{\bar{s}w}^u\right)_1 + \left(A_{\bar{s}w}^u\right)_2$) and air ($A_{\bar{s}a}^u$) react.

$$A_{\bar{s}}^u = A_{\bar{s}w}^u + A_{\bar{s}a}^u \tag{2.2}$$

Finally, in the dry fraction all particles are surrounded by air and also they are in contact with other particles. Again, if the contact area between particles is neglected then the total area of the dry fraction results from the addition of the area where air directly reacts plus the area of solids cut by line B-B', in the form

$$A^d = A_a^d + A_{\bar{s}a}^d$$

In this way, the total area for both the saturated and unsaturated fractions where the liquid phase reacts can be written as

$$A_w = A_w^s + A_{\bar{s}w}^s + A_w^u + A_{\bar{s}w}^u \tag{2.3}$$

and the total area where the air pressure reacts is

$$A_a = A_a^u + A_{\bar{s}a}^u + A_a^d + A_{\bar{s}a}^d = A - A_w \tag{2.4}$$

If a vertical force F is applied on the plane B-B', equilibrium is obtained from the reactions of each phase and the following equation can be written

$$F = \sigma_1 A = \sigma_{\bar{s}1} A_{\bar{s}} + u_a A_a + u_w A_w \tag{2.5}$$

where σ_1 is the total vertical stress and $\sigma_{\bar{s}1}$ is the vertical stress transmitted by the solid particles. The term $\sigma_{\bar{s}1} A_{\bar{s}}$ can also be written as $\sigma_{\bar{s}1} A_{\bar{s}} = \sigma_1' A$, where σ_1' represents the effective vertical stress which is related to the stress supported by the solid skeleton and therefore, it can also be related to the shear strength and the volumetric behavior of the material. Therefore, Eq. (2.5) can be rewritten as

$$\sigma_1 A = \sigma_1' A + u_a A_a + u_w A_w \qquad (2.6)$$

Manipulating Eqs. (2.4) and (2.6), the following relationship is obtained

$$\sigma_1' = (\sigma_1 - u_a) + (u_a - u_w)\left(\frac{A_w}{A}\right) \qquad (2.7)$$

where the term A_w/A represents Bishop's parameter χ. Similar equations to the one shown above have been proposed by Matyas and Radhakrishna [1] and Öberg and Sällfors [18]. On one side, Matyas and Radhakrishna [1] stated that A_w/A represents a measure of the saturated pore space and therefore this quantity depends mainly on the degree of saturation of the material. On the other side, Öberg [31] analyzed the value of the ratio A_w/A for a three dimensional ideal soil made of spheres in open and close packing. He observed that the value of this parameter was close to the degree of saturation (S_w) for both cases, especially when Sw $\geq$ 50%. He also reported that within this range, the difference between A_w/A and S_w was not larger than 20%. Accordingly, these authors concluded that, under certain conditions, the ratio A_w/A can be approximated to the degree of saturation of the soil, as its exact value is difficult to obtain in practice.

A more convenient way to write the quantity A_w/A is using common soil mechanics volumetric relationships. According to Eqs. (2.1) and (2.3), the ratio A_w/A can be expressed as

$$\frac{A_w}{A} = \frac{A^s}{A} + \frac{A_w^u + A_{sw}^u}{A} \qquad (2.8)$$

The term A^s/A represents the ratio of the saturated area (A^s) to the total area of the section (A). If pores are randomly distributed in a homogeneous isotropic material, it can be proven that the areas of water, air and solids appearing in a cross section of a representative elementary volume of the material, adequately represent the volumetric distribution of the phases [32]. Therefore, the areas corresponding to each phase can be related to their respective volumes, in the form

$$\frac{A^s}{A} = \frac{V^s}{V} = f^s \qquad (2.9)$$

where V^s and V represent the volume of the saturated fraction and the total volume of the material, respectively. In that sense, f^s is called the saturated fraction of the soil. Therefore, according to Eq. (2.1), the volume of the saturated fraction is obtained by adding the volume of solids completely surrounded by pores filled with water with the volume of those pores. The solids surrounded by water are called saturated solids. The saturated fraction is then obtained by dividing the saturated volume by the total volume of the material.

On the other hand, Eq. (2.2) can be rewritten as

$$A_{\tilde{s}}^u = A_{\tilde{s}w}^u + A_{\tilde{s}a}^u = A_{\tilde{s}}^u \frac{A_v^u}{A_v^u} = A_{\tilde{s}}^u \frac{A_w^u + A_a^u}{A_v^u} = A_{\tilde{s}}^u \frac{A_w^u}{A_v^u} + A_{\tilde{s}}^u \frac{A_a^u}{A_v^u}$$

where A_v^u represents the area of voids of the unsaturated fraction. Gathering together the terms involving air pressure on one side, and the terms involving water pressure on the other, the following relationships can be found:

$A_{\tilde{s}w}^u = A_{\tilde{s}}^u \left(A_w^u / A_v^u \right)$ and $A_{\tilde{s}a}^u = A_{\tilde{s}}^u \left(A_a^u / A_v^u \right)$. Then, using the relationship $A^u = A_{\tilde{s}}^u + A_w^u + A_a^u = A_{\tilde{s}}^u + A_v^u,$ the term $\left(A_w^u + A_{\tilde{s}w}^u \right) / A$ in Eq. (2.8) can be written as

$$\frac{\left(A_w^u + A_{\tilde{s}w}^u \right)}{A} = \frac{A_w^u \left(1 + A_{\tilde{s}}^u / A_v^u \right)}{A} - \frac{A_w^u}{A_v^u} \frac{A^u}{A} \tag{2.10}$$

Again, by considering the pores randomly distributed into a homogeneous isotropic material, areas and volumes can be related. Therefore, the above equation transforms into

$$\frac{A_w^u}{A_v^u} \frac{A^u}{A} = \frac{V_w^u}{V_v^u} \frac{V^u}{V} = S_w^u \left(f^u \right) \tag{2.11}$$

where V_w^u and V_v^u represent the volume of water and voids of the unsaturated fraction, respectively while V^u is the total volume of the unsaturated fraction, $S_w^u = V_w^u / V_v^u$ represents the degree of saturation of the unsaturated fraction and $f^u = V^u/V$ represents the unsaturated fraction of the soil. Therefore, the volume of the unsaturated fraction results from the addition of the volume of all solids surrounded by a combination of saturated and dry pores added by the volume of these pores. These solids are called unsaturated solids. Those pores filled with water but in contact with both saturated and unsaturated solids are considered part of the saturated fraction. This is so because they are part of the pores that surrounds a saturated solid. The unsaturated fraction of the soil is then represented by the unsaturated volume divided by the total volume of the material.

elements divided by the total volume of the sample. The volume of dry elements is obtained by adding the volume of dry solids with the volume of pores surrounding these solids. The dry solids are those particles exclusively surrounded by dry pores. The addition of the saturated, unsaturated and dry fractions results in unity

$$f^s + f^u + f^d = 1 \tag{2.12}$$

Finally, the degree of saturation of the unsaturated fraction is represented by the volume of water of the unsaturated fraction divided by the volume of voids of the same fraction.

Finally, by using Eqs. (2.7-2.11), the effective vertical stress can be written as

$$\sigma_1' = \overline{\sigma}_1 + s\left[f^s + S_w^u f^u\right] \tag{2.13}$$

where $\overline{\sigma}_1$ represents the net stress in the vertical direction. Accordingly, the tensorial form of the effective stress can be expressed as

$$\sigma_{ij}' = \sigma_{ij} - u_a \delta_{ij} + s\left[f^s + S_w^u f^u\right]\delta_{ij}$$

where δ_{ij} represents the delta of Kronecker. By multiplying the above relation by δ_{ij}, it becomes

$$p' = p - u_a + s\left[f^s + S_w^u f^u\right] = \overline{p} + \chi s \tag{2.14}$$

where p', p and $\overline{p}$ represent the mean effective stress, the mean total stress and the mean net stress, respectively. By comparing Eqs. (2.13) and (1.1) (Chapter 1), the value of Bishop's parameter χ can be found

$$\chi = \frac{A_w}{A} = f^s + S_w^u f^u \tag{2.15}$$

The variables f^s and S_w^u can also be written as a function of the global degree of saturation (S_w) and the void ratio (e) of the soil in combination with the variables $r_v^s = V_v^s/V_v$ and $r_3^s = V_3^s/V_3$, which represent the ratio of the volume of voids of the saturated fraction (V_v^s) to the total volume of voids of the soil sample (V_v) and the ratio of the volume of solids of the saturated fraction (V_3^s) to the total volume of solids of the soil sample (V_3), respectively,

$$S_w^u = \frac{V_w^u}{V_v^u} = \frac{V_w - V_v^s}{V_v - V_v^s} = \left(S_w - r_v^s\right)/\left(1 - r_v^s\right)$$

$$f^s = \frac{V^s}{V} = \frac{V_v^s + V_{\overline{s}}^s}{V} = \frac{V_v^s/V_{\overline{s}} + V_{\overline{s}}^s/V_{\overline{s}}}{1+e} = \frac{er_v^s + r_{\overline{s}}^s}{1+e}$$

where V_w represent the total volume of water of the sample. Then, Bishop's parameter χ can also be expressed as

$$\chi = f^s + S_w^u\left(1 - f^s\right) = \frac{er_v^s + r_{\overline{s}}^s}{1+e} + \left(\frac{S_w - r_v^s}{1 - r_v^s}\right)\left(1 - \frac{er_v^s + r_{\overline{s}}^s}{1+e}\right) \qquad (2.16)$$

That is to say, Bishop's parameter χ not only depends on the global degree of saturation of the soil but also on the global void ratio and the volume of voids and solids of the saturated fraction.

By applying the principles of thermodynamics and specifically that of enthalpy, Murray [29] proposed a coupled stress p'_c to normalize the shear strength of unsaturated soils subject to different suctions. The comparison with experimental results, however, showed some scatter and the results of saturated tests could not be included into the formulation. In his study, Murray [29] obtained the value of Bishop's parameter as $\chi = (1 + eS_w)/(1 + e)$, which is somehow similar to Eq. (2.16). The difference comes from the fact that Murray neglected the effect of air pressure on the solid particles.

Eq. (2.13) can also be written as

$$\sigma_1' = \overline{\sigma}_1 + \sigma_s^* = \sigma_1 - u_a + s\chi \qquad (2.17)$$

where $\sigma_s^* = s\chi = s\left[f^s + S_w^u f^u\right]$ is called the matric suction stress and represents that part of the effective stress generated by matric suction. In the case of saturated soils ($f^s = 1, f^u = 0, \chi = 1$), the effective stress becomes Terzaghi's effective stress. For a completely dry soil, as for example clean sand, ($f^s = 0$ and $S_w^u = S_w = 0$) the effective stress becomes the net stress. Finally, for the case of a soil with no saturated fraction, for example a poorly graded clean sand subject to a small suction where all water appears in the form of menisci ($f^s = 0, S_w^u = S_w$), the effective stress becomes Bishop's stress with $\chi = S_w$.

On the other hand, the equation of shear strength for unsaturated soils can be written as

$$\tau = \sigma'_n \tan \varphi = \left(\bar{\sigma}_n + \sigma^*_s \right) \tan \varphi \tag{2.18}$$

where σ'_n and $\bar{\sigma}_n$ represent the effective and the net normal stress, respectively. A similar equation has been proposed by Vanapalli, Fredlund, Pufahl and Clifton [21], even though the matric suction stress was written solely as a function of the degree of saturation and the residual degree of saturation. This equation states that the effect of suction on the strength of soil is simply the increment of the contact stress between particles produced by the presence of menisci of water, as assumed by Haines [33].

If the shear strength equation for unsaturated soils (Eq. 2.18) is derived with respect to soil suction, then the following relation is obtained

$$\frac{\partial \tau}{\partial s} = \tan \varphi \left\{ \left[f^s + S^u_w f^u \right] + s \left[\frac{\partial f^s}{\partial s} \left(1 - S^u_w \right) + \frac{\partial S^u_w}{\partial s} f^u \right] \right\}$$

with

$$\frac{\partial f^s}{\partial s} = \frac{(1+e)\left[\frac{\partial r^s_{\frac{s}{3}}}{\partial s} + r^s_v \frac{\partial e}{\partial s} + e \frac{\partial r^s_v}{\partial s} \right] - \left(r^s_{\frac{s}{3}} + e r^s_v \right) \frac{\partial e}{\partial s}}{(1+e)^2}$$

$$\frac{\partial S^u_w}{\partial s} = \frac{\left(1 - r^s_v \right) \left[\frac{\partial S_w}{\partial s} - \frac{\partial r^s_v}{\partial s} \right] + \left(S_w - r^s_v \right) \left(\frac{\partial r^s_v}{\partial s} \right)}{\left(1 - r^s_v \right)^2}$$

Then, the slope at the origin ($s = 0$ and $f^s = 1$), has the value

$$\left(\frac{\partial \tau}{\partial s} \right)_{s=0} = \tan \varphi$$

In other words, the initial slope of the increment of the shear strength with soil suction equals the internal friction angle. This result has also been experimentally observed by different researchers, for example: Escario, Jucá and Coppe [34] and Gan and Fredlund [35]. However, none of them attempted to explain this finding.

The equation of the failure surface on the plane of mean effective stress *versus* deviator stress can be written in the same form as for saturated soils

$$q = Mp' \tag{2.19}$$

where q represents the deviator stress and the slope of the failure surface M is given by

$$M = \frac{6 \sin \varphi}{3 - \sin \varphi}$$

Because of the hysteresis of the loading-unloading and drying-wetting curves, the behavior of unsaturated soils is influenced by the wetting-drying history of the soil. This influence has been observed by experiments carried out by Bishop and Blight [36], Allan and Sridharan [37], Nishimura, Hirabayashi, Fredlund and Gan [38] and Sivakumar and Wheeler [39]).

If the functions defining parameters f^s, f^u and S_w^u were known, then the matric suction stress could be expressed solely as a function of suction. That is why some of the equations proposed for parameter χ based exclusively on the value of suction or the degree of saturation and some properties of the SWRC, show certain convergence with the experimental results ([17, 21]).

The main drawback of the analysis shown above is that parameters f^s, f^u and S_w^u cannot be obtained from direct experimental procedures. For example, Klubertanz, Laloui, Vullict and Gachct [40] have been using the neutron tomography procedure and the image processing to study the flow of water and the strains of unsaturated materials. These images discriminate the solid, the water and the air phases in different sections of the material and the porous and solid structures of the sample can be observed using an image processor. Nevertheless, the resolution of this equipment is around 0.125 mm which means that this method cannot be used for silty or clayey soils and, only the images of coarse and medium sands have been generated using this technique. This inconvenient has been overcome by combining this method with the X ray tomography which allows a resolution of the order of microns. It can be expected that in the next future the combination of these techniques could provide experimental values for parameters f^s, f^u and S_w^u required to compute the effective stress of unsaturated soils.

In the meanwhile an alternative procedure for obtaining the values of these parameters is throughout a porous-solid model. A model of this type should be able to represent the structure of a real soil by considering the pore (PSD) and grain size distributions (GSD). Additionally, it should be able to reproduce the phenomenon of hydraulic hysteresis during wetting-drying cycles. A model of this type is described in the next chapter.

CHAPTER 3

The Porous-Solid Model

Abstract: Based on the study of the equilibrium of the particles of a soil sample subject to certain suction, in the previous chapter it was possible to establish an analytical expression for the value of Bishop′s parameter χ. This parameter can be written as function of the saturated fraction, the unsaturated fraction and the degree of saturation of the unsaturated fraction of the soil. However, the determination of these three parameters cannot be made from current experimental procedures. Therefore, a porous-solid model simulating the structure of the soil is proposed herein and used to determine these parameters. The data required to build the porous-solid model are the void ratio and the grain and pore size distributions.

Keywords: Bonds, Cavities, Distinct element method, Grain size distribution, Macropores, Mesopores, Micropores, Network porous models, Pore shrinkage, Pore size distribution, Porous-solid model, Random models, Sites, Soil structure, Soil-water retention curves.

3.1. INTRODUCTION

Only recently, it has been acknowledged that Bishop′s effective stress equation ($\sigma' = \bar{\sigma} + \chi s$) may lead to more realistic and simple constitutive models for unsaturated soils (see for example [22, 24, 25]). However, the problem of a proper determination of parameter χ still subsists as it has been experimentally recognized that the approximation $\chi \approx S_w$ is not satisfactory for most soils.

In the previous chapter, an analysis of stresses on the skeleton of an unsaturated soil showing a bimodal structure resulted in the effective stress equation for unsaturated materials (Eq. 2.13). Unfortunately, parameters f^s, f^u and S_w^u required for the determination of the effective stress cannot be obtained from current laboratory procedures.

An alternative procedure for the determination of parameters f^s, f^u and S_w^u is the use a porous-solid model able to simulate the distribution of water in the pores of soils and hence reproduce the soil-water retention curves (SWRCs).

Some simplified porous models have been already developed to study different phenomena such as capillary condensation and evaporation [41] and activated chemical absorption in heterogeneous surfaces [42]. Also, Fredlund and Xing [43] proposed an equation that defines the SWRC based on the pore size distribution (PSD). More recently Simms and Yanful [44] proposed a porous network that correctly simulates the SWRC, relative hydraulic conductivity, volume change and PSD. However, these latter models do not account for hysteresis. One way to include hysteresis and observe in detail the influence of water menisci on the deformation and volumetric behavior of unsaturated soils is by making use of a micromechanical model. This type of models is more complex because, besides simulating the porous structure, it also simulates the solid skeleton of the material and can include the phenomenon of shrinkage of macropores at drying. A model with these characteristics can be used to determine the values of parameters f^s, f^u and S_w^u required to compute the effective stress for unsaturated soils. Three of these models are described below.

3.2. DIFFERENT POROUS-SOLID MODELS

The accurate description of a real porous medium, such as soils, is a quite a complicated task, if only because they show millions or billions of pores per gram which sizes ranges from 0.01 to 500 micrometers. Another problem is the phenomenon of hysteresis. Already in 1929, Haines [28] postulated that the main drainage SWRC occurs at higher suctions than the main wetting curve, because the latter is controlled by the largest pores while the former, is controlled by the smallest. Additionally, when load or suction increases there is a reduction in the size of pores. The shrinkage of macropores with suction has been studied by Simms and Yanful [45] by analyzing the PSD change of a glacial till. They observed that the pore volume related to the pore size exhibits two crests, as shows Fig. (1). The first of these crests, the one located at approximately 0.1 μm, corresponds to the mesopores, *i.e.* those that maintain their size when suction increases. The other crest (located at approximately 5 μm) corresponds to the macropores, which shrink with increasing suction. Simms and Yanful [45] observed that for this particular soil, practically all macropores experienced a progressive shrinkage as suction increased. For suctions of the order of 2.5 MPa, practically all macropores had shrunk, and their size diminished to approximately the size of the non-collapsing pores. The same type of behavior was observed for other soils. Additionally, there is the shrinkage of pores with loading. Simms and Yanful [46] performed a series of PSD tests on different soils subject to different confining stresses. These results show a general trend for all cavities to reduce their size and displace their peak on the PSD curves to the left hand side with increasing confining stresses although the reduction in size of macropores is much larger than that of mesopores.

Accordingly, a simplified description of soils that captures the phenomena described above can be made with four elements: the macropores, the mesopores, the bonds and the solids. Both macropores and mesopores represent the cavities or sites of the porous media. These elements contain most of the volume of voids. The mesopores are those pores from medium to small size. The macropores are the largest pores in the soil and differ from the mesopores in that the former shrink with increasing suction or load. The bonds or throats are the elements that link together two cavities. These pores are the smallest in the porous media and are also called micropores. The volume contained by the bonds is negligible when compared to that of cavities. Finally, the solids are included in the spaces left by the pores and form the skeleton of the material. If an analogy is made between the porous structure of a soil and a building, then the rooms and corridors of the building represent the cavities while the doors and windows represent the bonds. Additionally, the solid structure of the building represents the skeleton of the soil.

Each one of these elements posses its own size distribution however, its spatial distribution is strongly correlated given the geometrical restrictions to be fulfilled. These interactions with their neighbors allow reproducing in a simplified manner the structure of soils. Therefore, a solid-porous network built with these elements can simulate the most important aspects of the wetting-drying phenomena, as for example the hydraulic hysteresis of the SWRC and the shrinkage of macropores. For this purpose, the model must comply with certain conditions to correctly describe the main phenomena of real soils. These conditions are:

a. Heterogeneity of sizes. Meaning that all elements (macropores, mesopores bonds and solids) have their own size distribution.

b. Compressibility of the network. This can be accomplished by allowing the shrinkage of macropores as suction or applied loads increase.

c. Size correlation between neighbors. Meaning that there is a statistical correlation between the sizes of the different elements meeting at a certain place such as cavities with bonds and cavities with solids.

d. Non-uniform connectivity, as the number of bonds converging at one site may change from site to site.

e. Geometrical restrictions, in order to guarantee that the bonds connecting to one site do not intersect one another.

f. Segregation between fine and coarse particles, as fine particles join together and appear in the form of peds or aggregates of different sizes.

g. Isolated clusters, as some pores remain inaccessible at wetting or drying.

The presence of isolated clusters is confirmed by the existence of the primary and secondary boundary curves. The primary boundary curves at wetting or drying are obtained from independent tests beginning with completely dry ($S_w = 0$) or fully saturated ($S_w = 1$) materials, respectively. However, when drying-wetting cycles are applied the degree of saturation never reaches these limits and the secondary boundary curves develop (see Fig. **2**). With further cycling, these curves remain unchanged [47]. This phenomenon has been called permanent hysteresis, and means that the relationship between the capillary pressure and the degree of saturation is not unique but depends on the history of the wetting-drying cycles [48]. It is clear that a real soil that has been exposed to wetting-drying cycles behaves according to the secondary boundary curves.

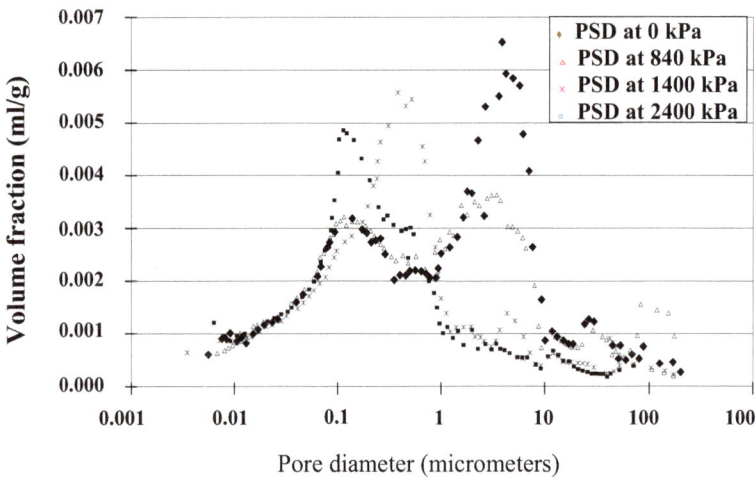

Fig. (1). PSD for a glacial till (after [45]).

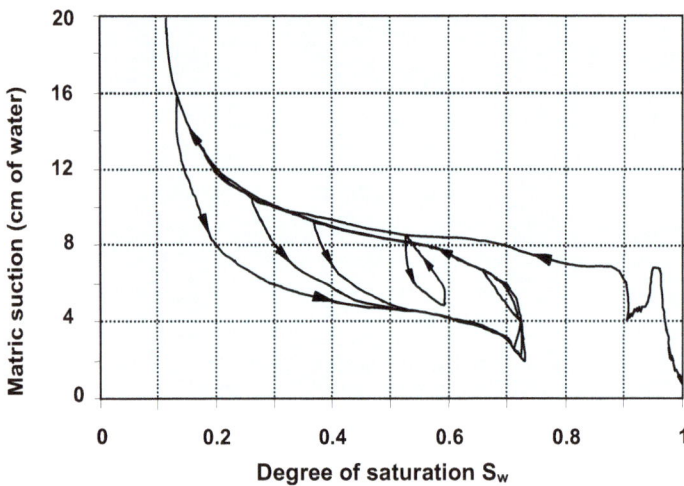

Fig. (2). Experimental SWRCs (after [49]).

3.2.1. Distinct Element Models

One of the first attempts to develop micromechanical models is due to Cundall and Strack [50] who developed the distinct element method (DEM). These authors considered the solids as disc shaped particles of different sizes. Knowing the spatial distribution (structure) of the solids and the external loads, contact stresses and deformations can be determined. At present, these models have evolved to more complex ones dealing with the micromechanics of unsaturated soils, among other applications. A model of this type has been developed by Gili and Alonso [51] and is briefly described below.

To properly simulate the behavior of unsaturated materials using a micromechanical model, menisci have to be added as new elements and the forces they introduce at the contact between solids need to be considered. In general, simple geometrical shapes as discs or spheres of different sizes are considered for solid particles (see Fig. 3). Then, the shape of the menisci and therefore the additional stresses on the solids solely depend on the arrangement or structure of the solids and the current water content (or suction). When a boundary condition is modified, the water content may change as a result of water transfer in the form of liquid or vapor. This transfer may occur along the surface of solids or through the pores. When this happens a new distribution of forces occurs and the spatial distribution of solids changes. Therefore the micromechanical model should include three different elements: the solid particles, the pores and the menisci of water. A model of this type can be very helpful in understanding the mechanisms involving the volumetric response of the material produced by a combination of load and suction. Also the effect of suction on the strength of unsaturated soils can be studied. In addition, micromechanical modeling can be used to verify some hypothesis made in elastoplastic constitutive modeling. The model can be two-dimensional (2D) or three-dimensional (3D); however a 3D model is required to realistically reproduce the forces induced by water menisci. This type of models is applicable mainly to sands and silts although it can also explain some features of the behavior of fine soils showing a structure made of aggregates or packets of particles. This type of structure can be obtained by compaction on the dry side.

In the micromechanical model, capillary forces are computed from the Young-Laplace equation for double capillary radii [28]

$$s = T_s \left(\frac{1}{r_1} + \frac{1}{r_2} \right) \tag{3.1}$$

where r_1 and r_2 represent the radii of the meniscus and T_s is the air-water interfacial tension. The contact force between two spherical grains of radii r

linked by a meniscus of water covering the surface of the grain with an angle 4θ measured from the center to the surface of the particle is according to Haines [33]

$$\Delta\sigma_n = \frac{\pi\, T_s(1 - 2\tan\theta)}{r(1 + \tan\theta)}$$

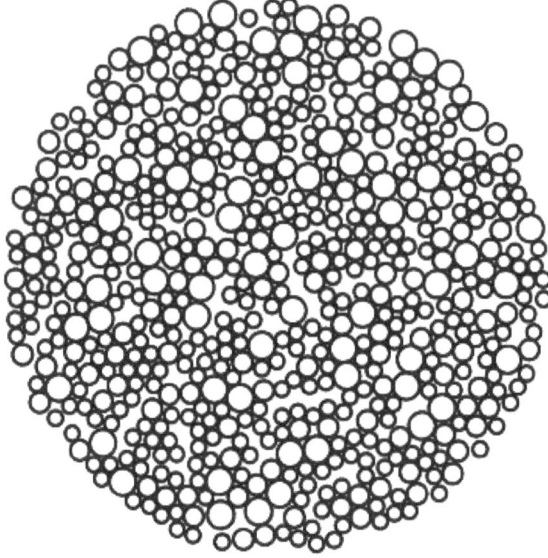

Fig. (3). DEM model considering 592 spherical particles (from [52] with permission from SMIG).

The current position of solids, pores and menisci is followed by three connectivity matrices: particle-menisci matrix, pore-menisci matrix and menisci-particle/pore matrix. The first two specify the menisci located on the periphery of a solid and the third identifies the two solids linked by a specific water meniscus. These matrices are required to establish the flow equations amongst different elements. All the transfer processes are described by a linear mass-flow rate equation of the type

$$\frac{\Delta M_{ij}}{\Delta t} = K_{ij}\left(p_j - p_i\right)$$

where ΔM_{ij} is the mass of a given species interchanged in a time increment Δt between entities i and j. p_i defines the pressure or concentration of a given species at entity i and K_{ij} is a generalized transfer coefficient which includes geometrical terms and constitutive flow parameters such as permeability and diffusivity. The types of flow considered by the model are: air-air, vapor-vapor, air (gas)-air (dissolved), water (vapor)-water-liquid, liquid-liquid, water-dissolved air. The first two occur exclusively in the pores, the following two occur between pores and menisci and the last two occur exclusively in the water menisci.

The equilibrium force and particle displacements are computed according to the procedure described by Cundall and Strack [50]. A 2D rheological model considering elastic, plastic and viscous units is used to simulate the contact between particles (see Fig. **4**). All individual particles must remain in equilibrium. At any contact between particles both normal (N) and shear stresses (T) are considered. The forces exerted by the menisci are considered to be normal to the tangent plane. The limiting shear force between particles is given by the condition $T \leq \mu N$ where μ is the contact friction coefficient.

This micromechanical model has been applied in the simulation of different stress paths, for example: isotropic loading at constant suction, collapse test including the effect of deviator stress, loading-wetting tests, loading-drying tests and wetting-drying cycles.

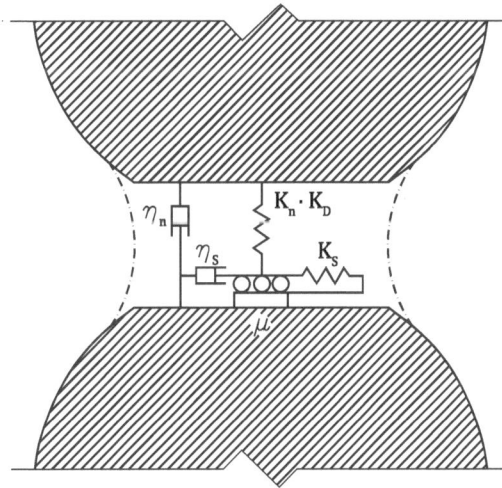

Fig. (4). Rheological model at particle contact (from [52] with permission from SMIG).

The simulation of the collapse test included the loading stage by increasing the mean net stress in several steps; from 2 kPa to 4 kPa, 6 kPa and finally 10 kPa. Fig. (**5**) shows the distribution of forces at contacts between particles at 2 kPa and 10 kPa. Observe that not only the distribution and intensity of the total forces change but also the shape of the sample modifies with loading. To simulate the collapse of the structure, suction has been reduced in two steps; initiating at 90kPa it has been reduced to 10 kPa and finally to 0 kPa. Fig. (**6**) shows the distribution of the total force at contacts when suction reaches 10 kPa. The sample reacts by reducing its volume and also modifying its shape.

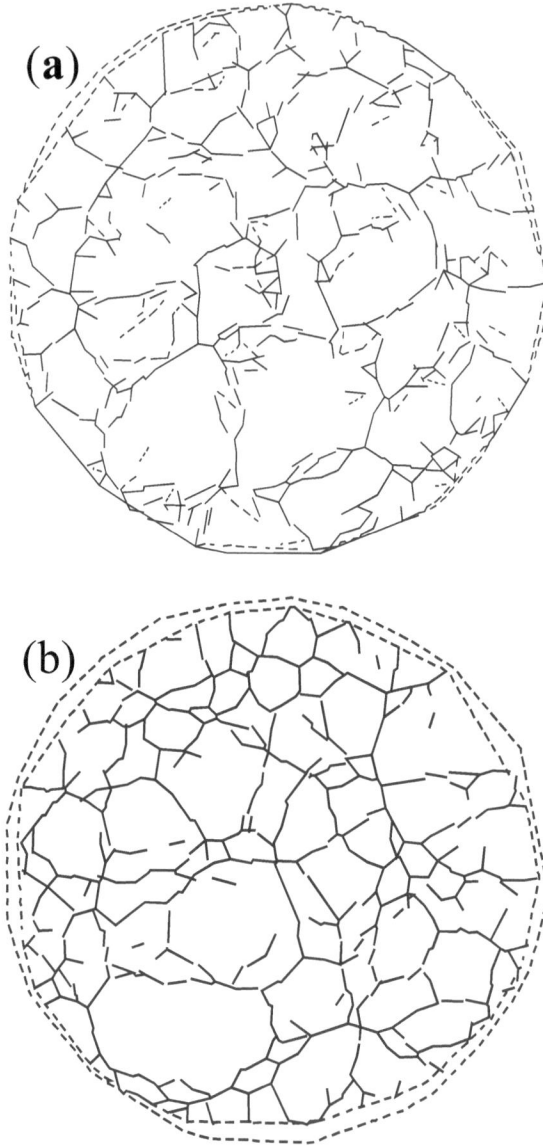

Fig. (5). Distribution of total forces at different net stress (**a**) 2 kPa, (**b**) 10 kPa (from [52] with permission from SMIG).

These type of simulations show in detail some important aspects of the behavior of soils for example: when loading is applied, the associated volumetric reduction is mainly the result of the destruction of larger pores giving rise to new smaller pores. Also the reduction of the potential of collapse with the increase in the intensity of confining stress can be reproduced on loose soils. This phenomenon occurs because the increase on the confining stress already causes an important volumetric reduction of the material.

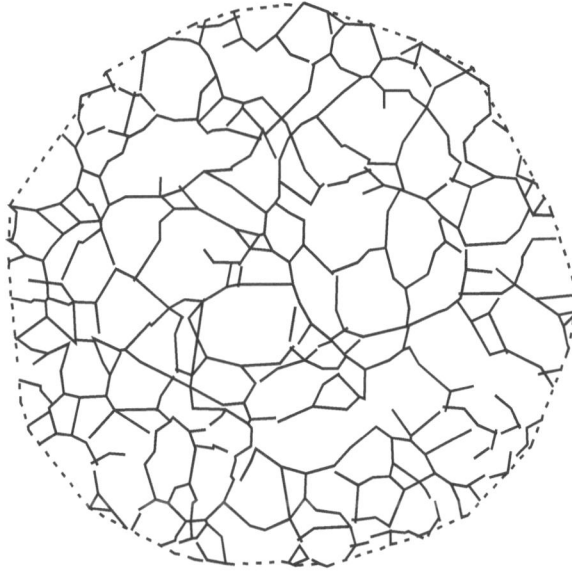

Fig. (6). Total force distribution when suction is reduced from 90 kPa to 10 kPa (from [52] with permission from SMIG).

Also with these simulations, the change in suction produced by an instantaneous loading can be studied as well as the volumetric response during drying-wetting cycles. Specially, the reduction of irreversible deformation with the number of drying-wetting cycles.

Due to computational constraints this type of models uses a limited number of solids, ranging between 1 000 to 10 000. The grain size distribution (GSD) and the water content are introduced as data whereas the PSD in the model is obtained by quantifying the number and size of the voids left by the solids when the initial target void ratio is reached. This procedure constitutes in fact one of the main disadvantages of this type of models as they cannot reproduce a particular PSD as it emerges from the random distribution of solids. Still solids can be arranged differently depending on the nature and formation process of the soil [53]. This means that, depending on the fabric process, a soil sample may show different PSDs even if both the GSD and the void ratio remain the same as is shown in Fig. (7). And because the SWRCs depend on the PSD of the material, a different set of SWRCs can be obtained for each fabric.

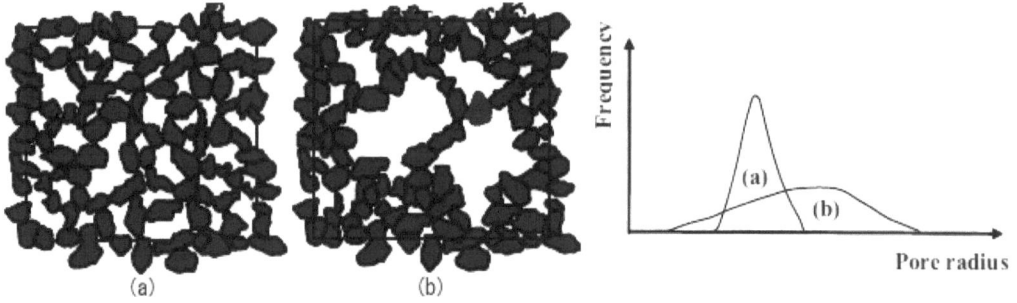

Fig. (7). Two different PSDs for the same GSD and void ratio (after [53]).

3.2.2. Random Models

Another type of models has been recently developed to include both the GSD and the PSD of the material [54]. These random models are built in a 2D or 3D grid made of squares or cubes. With the GSD, the PSD and the void ratio (*e*) of the material it is possible to define the number of solids and cavities of each size that need to be placed within a certain area or volume.

For the case of a 2D model a rectangular area with dimensions L in length and H in height is considered. In order to create an area with a number of elements that can be processed with ease by a common PC, the dimensions of the rectangular area are established by the following conditions: $20\ D_{max} < L < 250\ D_{min}$ and $H < 150\ D_{min}$ where D_{min} and D_{max} represent the diameter of the smallest pore and the largest solid to be placed on the rectangular area, respectively. These conditions were established in order to have an adequate number (not too small and not too large) of both the largest solids and the smallest pores. Once the area of the model has been defined, it is fully covered with squares whose sides represent $\sqrt{\pi}/2$ times the diameter of the smallest element in the model (D_{min}) equivalent to one pixel. Therefore, a total of $N = [L/D_{min}]x[H/D_{min}](4/\pi)$ pixels are placed within this area. These squares are called basic units and are used as a pattern to place the solids and cavities of the model. The number of basic units occupied by a solid or pore of certain size is established in such a way that all sizes are multiples of the basic unit. In the case of soils, the GSD usually shows larger sizes than the PSD although they may have some overlap. This means that the basic units represent the smallest cavities of the material.

The number of solids and cavities of each size are obtained from the GSD and the PSD of the soil, respectively, when plotted in the axis of volume fraction *versus* size. The volume fraction is defined as the ratio of the volume of solids (or cavities) which size ranges between certain limits to the total volume of solids (or cavities). To each volume fraction, the mean size within its range is assigned. The range of sizes for each volume fraction is selected in such a way that all mean sizes are multiples of the basic unit. As the void ratio represents the proportion between pores and solids, the product of the factor $e/(1+e)$ by the volume fraction of cavities divided by the volume of a single cavity corresponding to that size, results in the number of cavities of a specific size. In the same way, the product of the factor $1/(1+e)$ by the volume fraction of solids divided by the volume of a single solid corresponding to that size, represents the number of solids of that specific size.

Once the number of cavities and solids of each size has been determined, they are located in the model's area. In order to include all the required elements into the model's area, solids and cavities are placed at random following a size strategy which consists of placing these elements from the largest to the smallest. The location of solids and cavities initiates by randomly selecting a basic unit within the model's area where one of the largest solids is to be placed. Then, the basic units needed to represent the size of this solid are found using a polar searching procedure. This procedure consists of placing an origin of polar coordinates at the center of the selected basic unit and then turning around this origin with a constant radius to locate the closer adjacent elements that are available to generate the solid. The angle of rotation is gradually increased until a complete turn is made. If the required number of basic units is still not reached, the radius is increased. This process continues until the required number of basic units is found. All new elements formed during this process, in addition to complying with the established number of basic units, must also comply with a continuity condition. This condition establishes that a basic unit which is part of an element should have contact at least on one of their faces with another basic unit of this element. This condition allows the existence of different shapes for solids and cavities, but does not permit the generation of "strangled" elements *i.e.*, elements with basic units that have contact solely at one corner. Once all basic units constituting an element have been identified, they are assembled in a single element using a Boolean function. Additionally, all the basic units contained on its area are deleted from the list of available basic units. This procedure avoids the overlap between neighboring elements although they may have contact at the corners or sides of other elements. Once all the largest elements have been formed, the next smaller size elements (whether they are solids or cavities) are generated. This procedure continues until all solids and cavities have been located into the model's area. If at any location, the required number of basic units forming a continuous solid or pore cannot be found then another site is randomly selected until both conditions are fulfilled. Fig. (**8**) shows the flow diagram to locate and generate the solids and pores on the model's area.

With this procedure, the largest elements show more regular shapes than the smaller which tend to show irregular shapes. This occurs because the former are the first to be placed and have no restrictions in gathering adjacent basic units whereas the later are formed with the basic units left by the larger elements. This process ends with the elements one size larger than the basic unit as these last do not require the random location procedure.

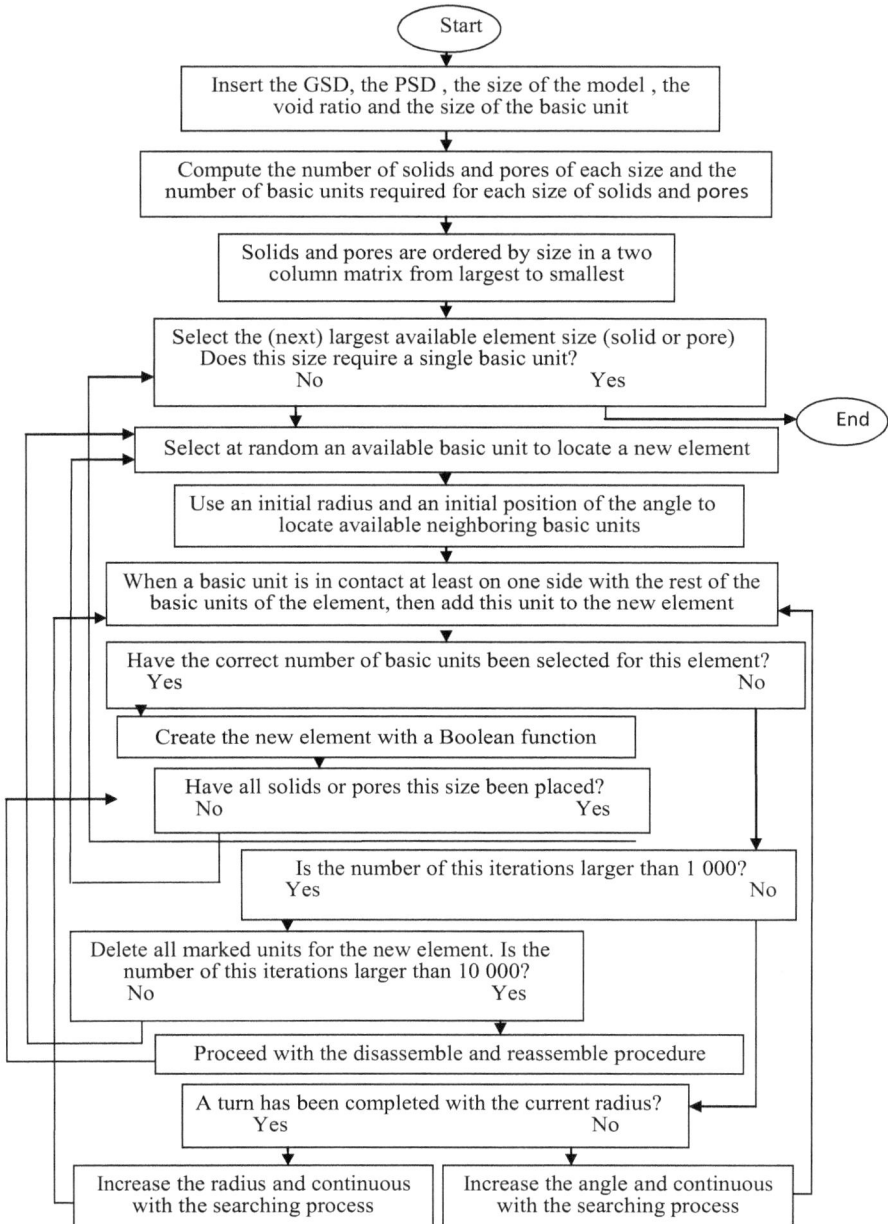

Fig. (8). Flow diagram for the construction of the random porous-solid model.

In most cases, not all elements of the smaller sizes can be located in the network because not enough groups with the required number of continuous basic units can be found. This usually happens with the elements one size larger than the basic unit because these are the last elements to be randomly placed in the network. These elements usually require groups of four or more basic units to be formed. To solve this problem, the following procedure has been adopted. First, the program identifies the groups with the larger number of basic units still available in the model and verifies if any of the neighboring solids or cavities around this group has additional adjacent basic units still available at their boundaries. If the program identifies the required number of basic units along the boundaries of these elements then it proceeds to rearrange them. To this purpose, the identified elements are disassembled into their basic units and then, the basic units required to increase the size of the group for the new element are liberated and substituted by the available basic units located at their boundaries. Finally all these elements are reassembled with their new basic units.

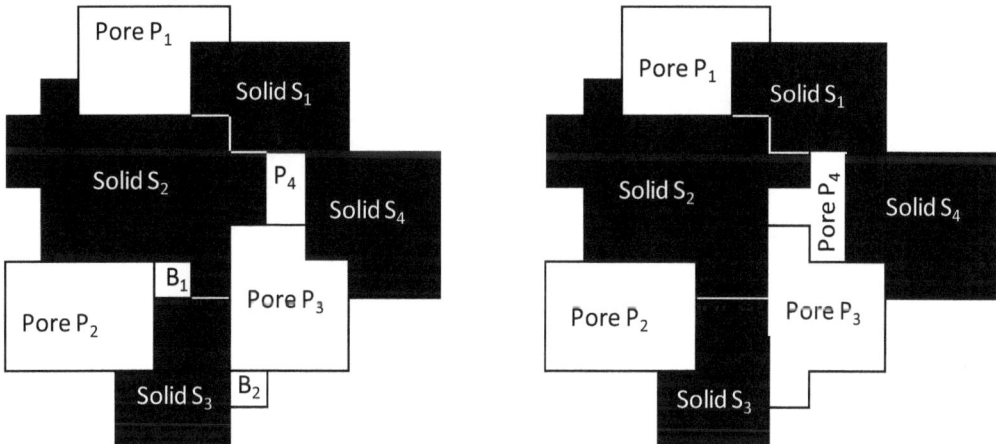

Fig. (9). Generating Pore P_4 by disassembling and reassembling Solid S_2 and Pore P_3.

This process is illustrated in Fig. (9). Suppose that pores made of four basic units constitute the elements one size larger than the smallest size represented by the basic unit. Consider that some of these pores have already been generated but still some more are required. However, no more continuous groups of four basic units can be found in the model's area. Consider also that at a certain location a continuous group of two basic units is found (pore P_4). Then the neighboring elements to this pore (Solid S_1, Solid S_2, Solid S_4 and Pore P_3) are analyzed to identify the basic units that are still available at their boundaries. In this case, elements S_2 and P_3 are selected as they are in contact with the basic units B_1 and B_2 still available. Next, S_2 and P_3 are disassembled to liberate two basic units that are in contact with pore P_4. Then S_2 and P_3 are reassembled by including the available basic units found at their boundaries. Finally, Pore P_4 is generated using the liberated units from its neighbors. With this procedure all cavities and solids can be inserted into the model's area.

When all solids and cavities have been located (see Fig. **10**), the bonds are included in the porous structure by linking neighboring cavities. A maximum and a minimum connectivity number are established for the model. This means that the number of bonds connected to each cavity can vary between these limits. These values can be established from the experimental results reported by Dullien [55] on a sandstone sample. Dullien [55] reported a connectivity varying from 2 to 10 with a mean value of 2.9 and very low frequencies for connectivity ranging between 7 and 10. By extrapolating these results to the case of soils, it can be said that their connectivity may vary from 2 to 6.

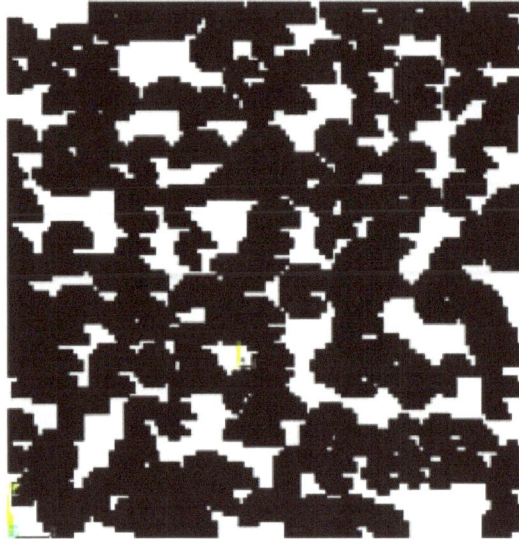

Fig. (10). Solid skeleton of the Vycor glass (in black).

The connectivity of pores in the model can be established by using the aforementioned polar search procedure. The search for the neighboring sites of a cavity starts with a radius slightly larger than the equivalent radius of the cavity being considered and is performed by increments of the polar angle. If the minimum connectivity number is not reached at the end of a complete turn, then the searching radius is slightly increased and the procedure continues until the connectivity reaches at least its minimum value at the end of a complete turn. This process can also be stopped when the number of connecting elements reaches the maximum established value. For the model discussed herein, a minimum connectivity of 2 and a maximum of 6 were considered according to the discussion above. When this process is concluded for all cavities, a size is randomly assigned for each bond. This size is obtained from the bond size distribution of the material. Because bonds are always smaller than the sites they connect to, the size of a bond is selected exclusively from those sizes smaller than the smallest of the two cavities to which it is connected to.

If the size distribution for both sites and bonds could be experimentally determined, then the porous model could predict both branches of the SWRC. Unfortunately at present, the experimental determination of the PSD for porous materials includes solely the size distribution of cavities because the volume of bonds is, in general, negligible compared with that of cavities. Therefore, to overcome this lack of information, the shape of the curve of the size distribution of bonds is considered to be similar to that of cavities while the size of bonds is obtained by adjusting the numerical drying SWRC with the experimental one. This is equivalent to horizontally displace the size distribution curve of bonds in the axis of size until the best fit between the numerical and the experimental drying curve is obtained. The drying curve is the one to be fitted because it is mainly dependent on the size distribution of bonds as aforementioned. The adjustment of the size of bonds practically does not affect the wetting curve which is mainly dependent on the size distribution of cavities. In addition, as bonds are the last elements to be placed in the model, their size can be modified once solids and sites have been located. This facilitates the fitting process for the drying curve. The necessity of fitting the drying SWRC represents an important drawback to the model which could only be avoided by supposing a value for the ratio between the size of cavities and bonds as both curves are considered to show similar shapes.

Once all sites, bonds and solids have been distributed in the model, it is possible to simulate the main wetting SWRC as well as the scanning curves. To achieve this, each site is transformed into a circle of the same area and placed at the centroid of the irregular pore.

To simulate the main drying curve it is assumed that all pores are initially saturated and that suction equals zero. Then suction is increased by steps and the critical radius (R_c) determined. This radius represents the lower limit of the size of pores able to drain during a drying process, and is given by the Young-Laplace equation [34] in the form of:

$$R_c = \frac{2\,T_s \cos\theta}{s} \tag{3.2}$$

where θ is the water-soil mineral contact angle. In general, it is considered that for the case of water with most soil minerals $\theta = 0$. According to Eq. (3.2), all pores with radius R complying with the condition $R \geq R_c$ will be able to dry. Therefore the model can identify all pores able to dry at the current value of suction.

Once all pores which size is larger or equal to the critical radius have been identified, they are considered as potentially active, meaning that they can dry if they comply with the continuous path principle. This principle states that a pore can dry only if it is connected to a continuous path of previously dried pores that reaches the boundary of the model where the bulk of gas is present. This type of pores is called active pores. This means that all boundary elements connected to the bulk of fluid (air or water) are already active pores whereas those elements located at the interior of the porous model need to be tested for this condition.

The same procedure can be followed to determine the main wetting curve except that in this case, all pores are initially dry, suction has a very large initial value which reduces with every step and the potentially active pores are those complying with the condition $R < R_c$.

For the scanning curves, the initial drying and wetting stage is the same up to the point where the inversion takes place. When inversion occurs, only those pores that have been activated will be able to deactivate. This means that solely those pores that have been dried during a drying process are able to saturate when inversion takes place. Correspondingly, only those pores that saturate during a wetting process are able to dry after the inversion.

The procedure to build the porous-solid model allows the generation of different structures for the same material meaning that even if the GSD and the void ratio are maintained constant, different PSDs can be generated. This characteristic is not considered by the porous models based on the DEM described earlier. With this model it is possible to study the effect of the structure of soils on the SWRCs. At its present stage this model can only roughly simulate the structure of soils. Refinements including a 3D version with larger number of solids, pores and size categories may result in better predictions. It is also possible to isolate the solids from the porous structure of the soil and include rheological models at the contacts between solids. This may allow simulating different phenomena as the volumetric deformation of the solid skeleton by loading or suction increase and the resulting variation of the PSD. Also the strength of the material at different suctions can be simulated in the same way as DEM models do.

One of the main restrictions to this type of models is the number of elements that a computer is able to process efficiently. In its present version, the model can deal with 40 000 basic units. The size of the scratch memory of the computer is the main constraint to increase the number of elements in the model. This constraint imposes additional restrictions related to the range of pore and solid sizes that the model is able to consider. For example, if the model considers a large variation of sizes for solids and pores, the number of basic units required for the largest elements will be so big and the number of these elements will be so small that the SWRCs may show abrupt changes. Therefore, the range of sizes that the model is able to consider is of about one order of magnitude for both the GSD and the PSD. This also restricts the type of materials that can be considered by the model. For example, at present only the structure of clean sands or clean silts can be simulated.

The computing time of the model greatly depends on the number of basic units and the number of elements of the solid-porous structure. A model made of 10 000 basic units with six different sizes of solids and pores takes around three hours of processing in a common PC to simulate the main SWRCs.

Figs. (**10** and **11**) represent the solid and porous structure of the Vycor glass obtained from its PSD. The squares in Fig. (**11**) represent the cavities and the lines represent the bonds. In this case a minimum of 2 and a maximum of 6 bonds were assigned to each cavity. Fig. (**12**) shows a 3D view of the porous structure of Vycor glass. Finally, Fig. (**13**) shows the numerical and experimental main SWRCs as well as the numerical and experimental comparison for two wetting scanning curves.

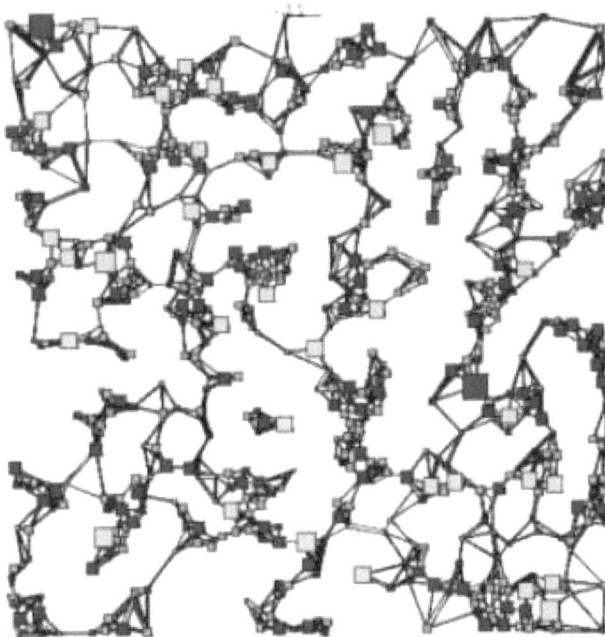

Fig. (11). Porous structure of the Vycor glass.

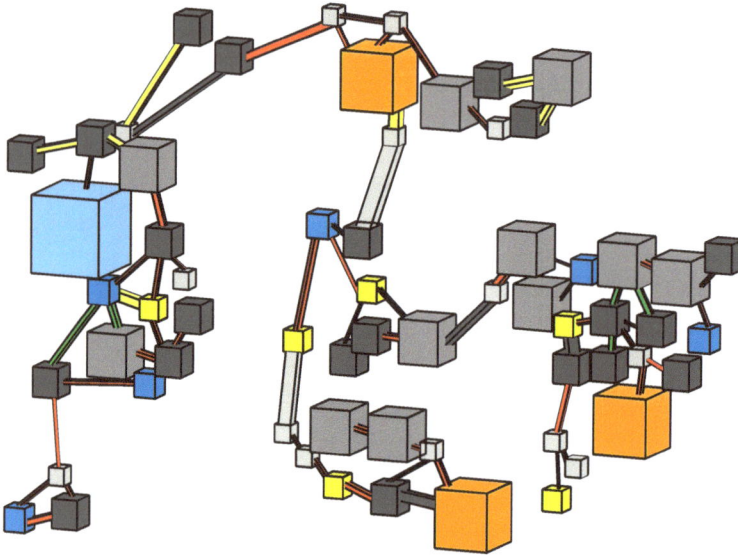

Fig. (12). 3D view of the porous structure of the Vycor glass.

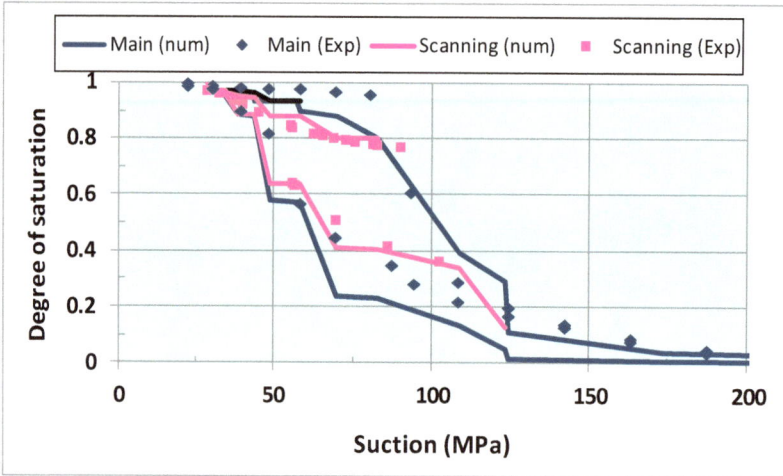

Fig. (13). Numerical and experimental main SWRCs and two drying scanning curves.

3.2.3. Network Models

Another type of porous-solid models is that developed on regular networks made of nodes and connectors. The nodes represent the cavities and the connectors are the bonds. Solids are included in the spaces between these elements. The simplest porous-solid network models are those built on the plane where the cavities are represented by circles and the bonds by rectangles. In a 3D configuration, the cavities are represented by spheres and the bonds by cylinders. The connectivity (*C*) represents the number of bonds meeting at a certain site and may have a constant value or it may vary from site to site. In this last case, it is sufficient to consider a certain number of bonds of size zero.

Due to the presence of elements with different sizes, the model has to comply with a construction principle representing a geometrical restriction. This principle ensures that the porous network is physically possible. This principle states that in order to avoid the intersection of two concurrent bonds (*i.e.* those that form an angle of 90° at the intersection with the concurrent site with radius R_s) with radii R_{B1} and R_{B2}, the following condition should be fulfilled

$$\sqrt{R_{B1}^2 + R_{B2}^2} \leq R_S$$

An important parameter that affects the morphology of the porous networks is the overlap between the sizes of bonds and sites. The larger the overlap, the more complex becomes to generate the network. In fact there is a limit on the value of overlap that allows the generation of the network without violating the construction principle. In general, when there is a considerable overlap and a small but well differentiated number of sizes, the phenomenon of segregation appears. This phenomenon groups together elements of the same size in different zones of the network. In this way, small sites gather around and are surrounded by bigger ones. This property allows the modeling of the structure of soils in a simplified way. A model of this type is described in the next section.

3.3. THE NETWORK MODEL

Similar to random models, network models include all four elements (macropores, mesopores, bonds and solids) distributed on a regular network. The solid-porous network is generated following the Monte Carlo procedure. Positions for mesopores and bonds are initially assigned at random to completely fill the network. Thereafter the construction principle is verified at every node. If the principle is violated at a certain location, an exchange between sites and/or bonds from another location determined at random is simulated. If the number of infractions decreases with this exchange, then it is granted. This process continues till no infraction in the construction principle subsists.

Once this process is completed, then macropores are placed by substituting the required number of mesopores. Macropores are also placed at random, merely verifying that the site to be replaced is smaller than the substituting macropore. If this is not the case, then another site is selected at random until this condition is fulfilled. This procedure ensures the observance of the construction principle.

In the same way, solids are also placed at random but following a size strategy. This strategy ensures that the size of solids is related to the mean size of its surrounding pores. For example, according to Taylor [56], the pore size D between three spheres of the same size d in close arrangement is $D = d/6.5$; hence, here it is considered that the size of a solid the size of gravel or sand to be placed in a certain position of the network must be larger than three times the mean size of their surrounding pores, that is to say $3D \leq d$. On the other hand, clay particles exhibit a flat shape and, consequently, the aggregation of several particles may generate pores larger than their equivalent radii. Therefore, considering that fine particles associate in structures of four or more elements, with a void ratio larger or equal to two, it can be established that the maximum pore size D should be smaller than $d_e \leq 0.5D$, where d_e represents the equivalent diameter of clay particles. Silt particles lay between sand and clay, and therefore, their size lie between the limits $0.5\,D \leq d \leq 3D$.

The distribution of solids in the pore network starts by specifying the size of the solid (gravel-sand, silt or clay) to be set in a particular place of the network. This can be accomplished by retrieving the maximum and minimum sizes of pores surrounding that specific place. In general, pores of the same size can be found in different areas of the network when the construction principle is being observed. Then a solid is picked up at random. If it does not comply with the required size range, another solid is picked up. If no more solids of the required size are available, then the range of possible sizes is enlarged until a solid can be found. A certain percentage of infractions among the total number of solids is accepted since, at the boundaries between fine and coarse materials, the presence of pores of very different sizes may lead to geometrical restrictions that no solid can satisfy. With these rules it is possible to build up a network that may approximately simulate the real structure of soils for both the solid and porous structure.

This model considers that the size distribution of all elements in the network (macropores, mesopores, bonds and solids) follow a logarithmic normal distribution, meaning that only two data are required to define each one of these distributions: the mean size ($\bar{R}$) and the standard deviation (δ). These parameters can be obtained from the PSD and the GSD of the material. It is also possible to use double or triple logarithmic normal distributions to better reproduce the PSD or the GSD of the soil. In such a case, additional parameters relating the proportion between the volumes of the different distributions are required. One additional parameter is required for a double normal distribution and two additional parameters are required for a triple normal distribution. This parameter is called here proportional volume factor, P_{vf}.

The model can be built in two or three dimensions and in different sizes. However, as all elements in the network are placed at random, it implies that different networks could be built with the same data and hence, the principle of uniqueness of results could not be respected. This problem can be overcome when sufficiently large networks are used. In fact, this is a problem also faced during experimental procedures as the soil sample should be large enough to represent the properties of a specific mass of soil. Therefore, large networks are preferred to small ones as they lead to more realistic and consistent results especially when the overlap between bonds and sites is large. In opposition, this type of networks increases significantly the computing time. In fact, an important restriction for the development of random networks is the computer time and the requirements in memory size. For example, the biggest network that a PC can manage with ease is a 1000 x 1000 plane network or a 100 x 100 x 100 3D network. If an important overlap exists between sites and bonds, the required computing time to construct the network in a common PC is around 8 hours.

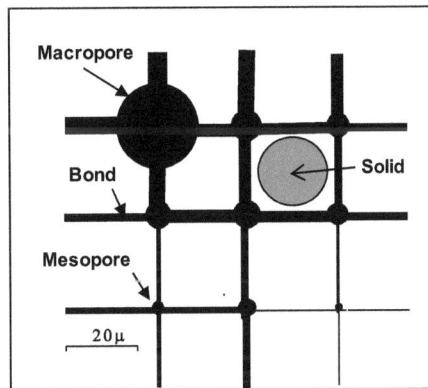

Fig. (14). The elements of the porous-solid network: macropores, mesopores, bonds and solids.

Fig. (**14**) shows a portion of a network built according to the procedure described above. It indicates the four different elements included in the network: macropores, mesopores, bonds and solids. The upper left-hand corner of the network represents a portion of soil tending towards the coarse fraction where sand particles are accommodated. The lower right-hand corner represents a portion of soil tending towards the fine fraction where clay particles are set. Between these two groups a transition of silt is present and fills most of the network. Notice that the distribution of solids and pores is made on a regular grid for practical purpose, however the size of solids is not directly linked to the size of the space between the pores but rather to the pores size, as mentioned before.

The length of bonds is considered as constant throughout the network. Because bonds are basically windows connecting two cavities they are considered to be very thin, therefore their length is considered to be similar to the size of the smallest cavities. In any case, this parameter only slightly affects the results provided by the porous-solid model as the volume of bonds is negligible when compared with that of sites as aforementioned.

In addition, when the distribution of all elements is completed, it is necessary to verify that the ratio between void and solids volume corresponds to the void ratio of the soil. Because clay particles generally show a flat shape and the hydrometer technique reports equivalent radii that have little rapport with their volume, a parameter called shape factor S_f is introduced in order to adjust the solid volume to its real value. This parameter tends towards unity for clean rounded sand but decreases dramatically when the fine fraction of the soil increases, especially for clays with high specific surface.

The model described above seems to be more appropriate for coarse materials where solids and pores show more or less spheroid shapes. In the specific case of clays, the solid particles are rather flat and the shape of pores largely varies from spheroids to slots [57]. In this last case, the size of bonds may be of the same order as that of sites and therefore the model proposed herein is probably not the most appropriate. In addition, in the case of clays subject to small water contents, interparticle forces are dominated by the van der Wals interactions [58] which are not considered in this model. In any case, only the comparison between experimental and theoretical results would give light on this issue.

3.4. MECHANISMS OF WETTING AND DRYING

Consider a porous network made of sites and bonds showing certain overlap in their size distributions. Consider also that no isolated clusters develop during the wetting and drainage phenomena. Therefore, in the first stage of its development, the proposed model only attempts to simulate the primary boundary curves. Finally, assume that pores fill and empty according to Young-Laplace's equation given by relationship 3.2.

Hence, when all pores are filled with water (*i.e.* suction is nil) and gas is forced into them by increasing suction, the first elements to drain are the largest, while the smaller only drain when suction is further increased. This means that bonds control the drainage process. On the other hand, when a dry soil is subject to a wetting path and a small decrement of suction is applied, the first elements to be filled in with liquid are the smallest, while the larger can only be filled when further reductions in the value of suction are applied. This means that cavities control the wetting process. Under these considerations, the conditions for the drainage and imbibition of pores can be stated and the primary boundary curves obtained.

3.4.1. Main Drying Curve

Consider that a soil sample is undergoing a drying process where some sites and bonds still remain saturated. The required conditions for a bond to dry are the following: a) gas has penetrated at least one of the bonds surrounding the two sites linked by the bond being considered (given that sites are always larger than bonds, the former do not show any restriction regarding drainage, once one of their connecting bonds has been invaded) and b) gas should be able to penetrate the bond under consideration, meaning that the bond is larger or equal to the critical radius given by Eq. 3.2.

On the other hand, the conditions for a site to be filled with gas are as follows: a) at least one bond of the contiguous sites has already been drained. This means that at least one site contiguous to the one under consideration has been already filled with gas as sites show no opposition to be drained once one of their bonds has been drained. b) Gas should be able to fill the bond that links the site already filled with gas with the one being considered. In other words, this bond should be smaller or equal to the critical size.

3.4.2. Main Wetting Curve

Consider now that the soil is undergoing a wetting process with some sites and bonds already saturated. For a bond to be saturated, the following conditions should be met a) at least one bond contiguous to the one being considered has to be saturated and b) water must be able to penetrate the site linking the saturated bond with the one under consideration, that is to say this adjacent site has to be smaller than the critical size.

For a site to be liquid filled, the following conditions must be fulfilled: a) at least one site contiguous to the one being considered is already saturated and b) the site under consideration can be invaded by water; in other words, it should be smaller than the critical size.

The degree of saturation can be easily obtained by dividing the current volume of sites and bonds filled with water by the total volume of pores.

3.4.3. Secondary and Scanning Curves

Once the primary boundary curves have been defined, the secondary curves can be easily obtained by considering that a certain number of elements remain inaccessible. This is readily done with the aid of a cluster coefficient C_c, defined as the ratio of the volume of pores belonging to closed clusters to the total volume of pores. One value is required for the maximum degree of saturation reached during wetting and another for the residual degree of saturation reached during drying. In this way, maximum and minimum values for the degree of saturation are given to the primary curves so they become the secondary boundary curves.

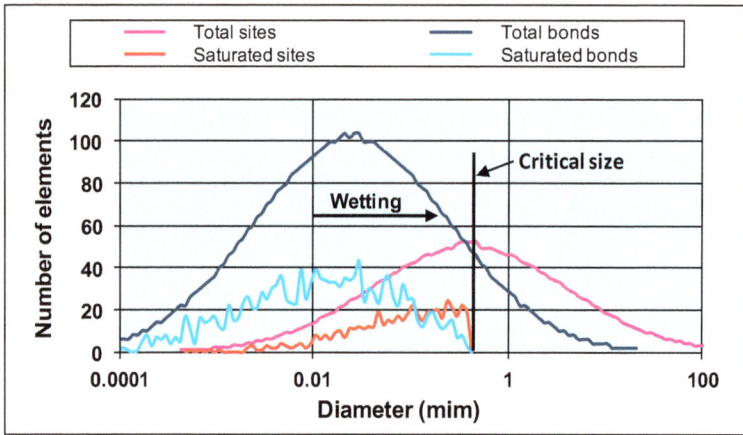

(a)

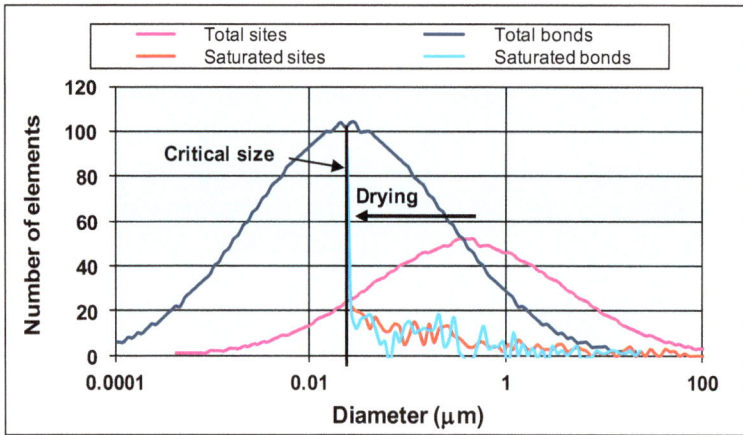

(b)

Fig. (15). Comparison between the saturated and total distributions for sites and bonds during a (**a**) wetting path and (**b**) drying path.

In case of inversions during a wetting or drying path, the conditions required for the wetting or drainage of sites and bonds remain the same as for the boundary curves. The only change results from the number of sites and bonds able to saturate or drain. For example, consider that an inversion arises during a wetting path with the distribution of sites and bonds already saturated shown in Fig. (**15a**).

Therefore, during the drainage stage, only those sites and bonds that have already been filled with water will be able to drain. Fig. (**15b**) shows the distribution of sites and bonds filled with gas during a drying path. Hence, when an inversion occurs, only those sites and bonds already filled with gas can saturate. A flow diagram to build a computational network model is shown in Fig. (**16**).

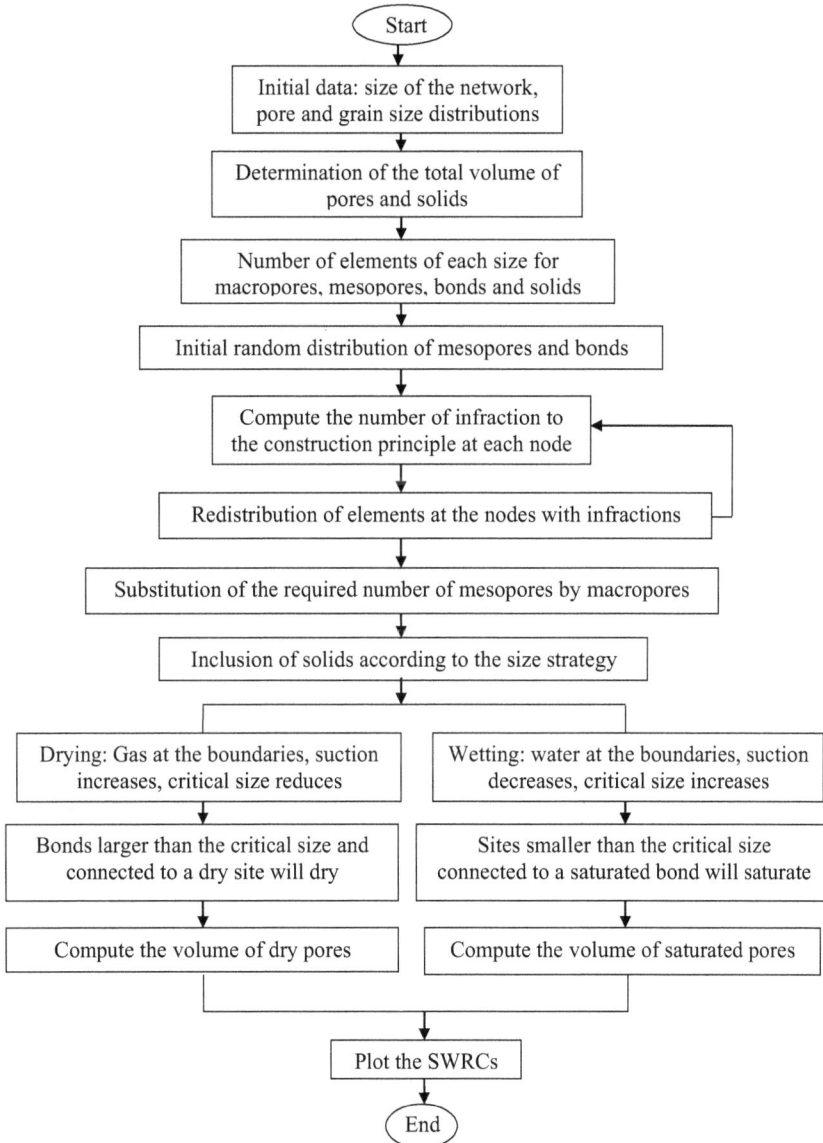

Fig. (16). Flow diagram to build a computational network model.

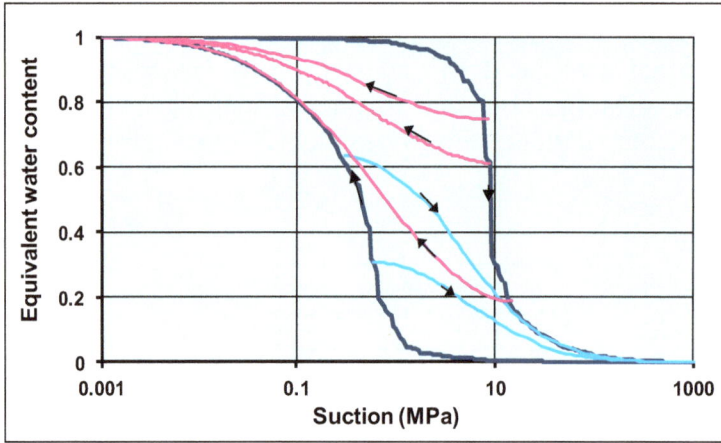

Fig. (17). Boundary and scanning curves obtained from the computational network porous-solid model.

Fig. (**17**) shows the wetting and drying boundary curves as well as some scanning curves obtained with the computational network porous-solid model. These numerical results are similar to the experimental curves shown in Fig. (**2**). One of the main differences between Figs. (**2 and 17**) is that the numerical scanning curves become asymptotic to the secondary boundary curves, while the experimental results show that the scanning curves reach the boundary curves at a point not far from the inversion point. This behavior needs to be studied further to improve the results of the model.

An option to overcome the constraints related to computing time and memory size in network models is the use of probabilistic network models [59]. In that case, the boundary curves can be established from the probability of a pore of certain size to be drained or filled with water when subjected to certain suction. A model of this type is developed in the next chapter.

CHAPTER 4

The Probabilistic Porous-Solid Model

Abstract: In the previous chapter, a computational network porous-solid model was developed to simulate the hydraulic behavior of unsaturated soils. However, important computational constraints make this model unpractical. In this chapter, a probabilistic porous-solid model is developed to overcome these constraints. The probabilistic model is an alternative to the use of computational network models and shows important advantages. This model is built by analyzing the probability of a certain pore to be filled or remain filled with water during a wetting or drying process, respectively. The numerical results of the probabilistic model are compared with those of the computational network model showing only slight differences. Then the model is validated by doing some numerical and experimental comparisons. Finally, a parametric analysis is presented.

Keywords: Basic unit, Bishop's parameter, Bonds, Cavities, Degree of saturation of the unsaturated fraction, Dry fraction, Hydro-mechanical coupling, Macropores, Mesopores, Micropores, Network models, Porosimetry tests, Probabilistic model, Relative volume, Retention curves, Saturated fraction, Solids, Unsaturated fraction.

4.1. INTRODUCTION

Recently, Bishop's stress equation has been used for the development of simpler and more realistic constitutive models for unsaturated soils ([22, 24, 25]) not only because it can estimate approximately the strength of soils but also because it takes into account the hydro-mechanical coupling observed in unsaturated soils. This phenomenon becomes evident by the fact that the degree of saturation affects the stiffness and strength of soil samples subject to the same suction. Part of this phenomenon can be related to the hysteresis of the soil-water retention curves (SWRCs) as, for a single value of suction a large range of values of the degree of saturation are possible. Another part can be related to the volumetric deformation of the sample during loading or suction increase, the so called hydro-mechanical coupling, as this in turn affects the SWRCs.

The analysis presented in Chapter 2 shows that Bishop´s effective stress equation for unsaturated soils (Eq. (1.1), Chapter 1) can be expressed as in Eq. (2.13) with parameter χ defined as in Eq. (2.15). According to this last equation, parameter χ

Eduardo Rojas

depends not only on the degree of saturation (S_w) of the sample, but also on the void ratio and the structure of the soil as experimentally observed by Bishop and Donald [6]. The main problem with the use of Bishop's equation lies precisely on the determination of parameter χ. In this chapter, a probabilistic porous-solid model is developed for the determination of this parameter using the SWRCs of the material as data.

4.2. THE PROBABILISTIC MODEL

Based on the framework of the computational network model, it is possible to develop a porous-solid probabilistic model [59]. The model is based on the concept of basic unit which allows the introduction of the solid phase resulting in a porous-solid model that can be used to determine the current effective stress in an unsaturated material. Initially, a basic unit for sites and bonds is defined and the equations for the main boundary curves at wetting and drying are obtained.

The procedure to develop the probabilistic model is as follows: First, an infinite 2D or 3D network made of macropores, mesopores, bonds and solids is considered. Thereafter, the conditions for a pore (cavity or bond) to drain or saturate during a drying or wetting process are established. Then, based on the size distribution of each element, it is possible to write the above conditions in the form of probability equations. These equations can then be simultaneously solved and the probability for a pore of a certain size to drain or saturate during a drying or wetting process can be determined. Subsequently, it is possible to establish a ratio between the dried and saturated pores and thus obtain the degree of saturation of the material to finally plot the SWRCs in wetting and drying.

This process requires knowledge on the distributions of the relative volumes of cavities (V_{RS}) and bonds (V_{RB}) as a function of their size. The relative volume is defined as the volume of the elements of certain size, divided by their total volume. These distributions can be obtained from the results of porosimetry tests. Once these distributions are known, it is possible to define the relative volume of cavities (macropores and mesopores) $S(R_C)$ and bonds $B(R_C)$ smaller or equal to the critical radius R_C in the form,

$$S(R_C) = \int_0^{R_C} V_{RS}(R)\,dR \tag{4.1}$$

$$B(R_C) = \int_0^{R_C} V_{RB}(R)\,dR \tag{4.2}$$

When these integrals are solved for the full range of sizes the result is unity, which means that these functions in fact represent the distribution of probabilities for sites and bonds, respectively. These functions are represented in Fig. (**1a** and **b**) for wetting and drying paths, respectively.

Using the above equations it is possible to determine the volume of pores of certain size. For example, if V_s represents the total volume of cavities, the product $V_S\,S(R_C) = V_S \int_0^{R_C} V_{RS}(R)\,dR$ represents the volume of cavities which sizes range from zero to R_C.

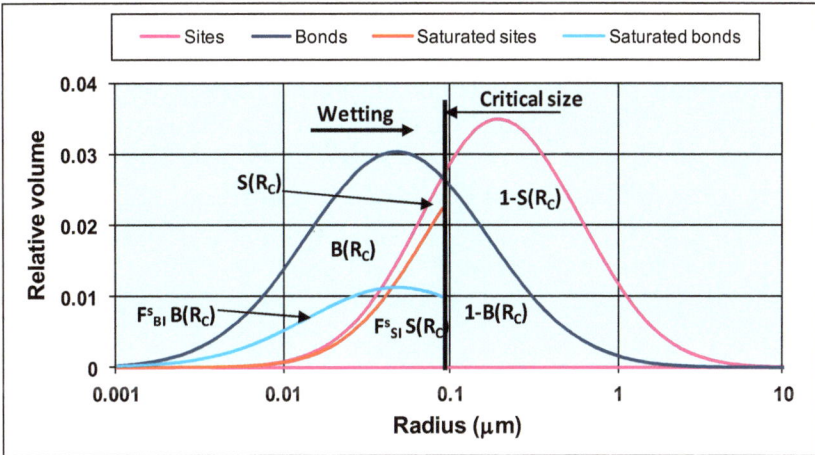

(a)

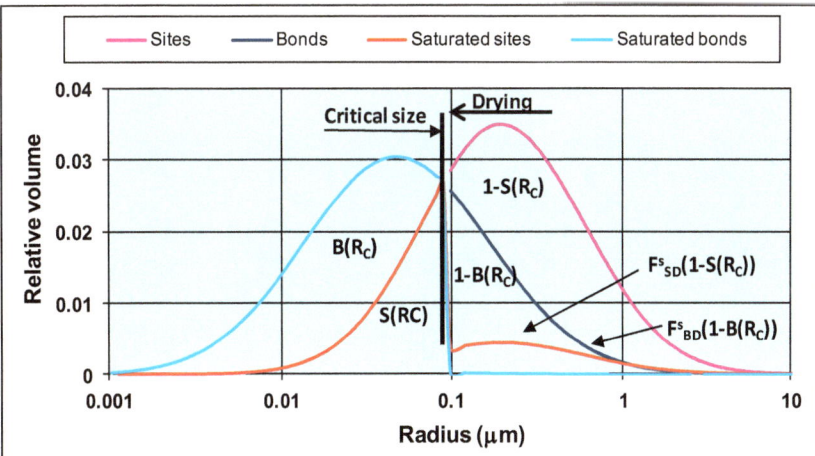

(b)

Fig. (1). Relative volume distribution for saturated cavities and bonds at (**a**) wetting and (**b**) drying.

4.3. MAIN WETTING CURVE

Consider a dry soil subject to a wetting path under controlled suction. During this process, some air bubbles may remain trapped within the irregularities of solid grains. Additionally, some pores may remain dry because, for example, all bonds connecting to a cavity may saturate before the cavity can fill with water because it is larger than the current critical size. However, for simplicity it is initially considered that all pores can saturate. That is to say, the equations for the primary SWRC at wetting and drying are initially developed.

Fig. (**2a**) shows the basic unit for cavities in a 2D porous network. It consists of a central cavity connected to four concurrent bonds each of which is connected to an external cavity. Consider that this basic unit is initially dry and subject to a wetting process. By inspecting this unit it can be established that the cavity can saturate only if the following two conditions are fulfilled: a) its radius is smaller than the current critical radius so, water can intrude the cavity and b) at least one bond connected to this cavity is already saturated and connected to the bulk of water.

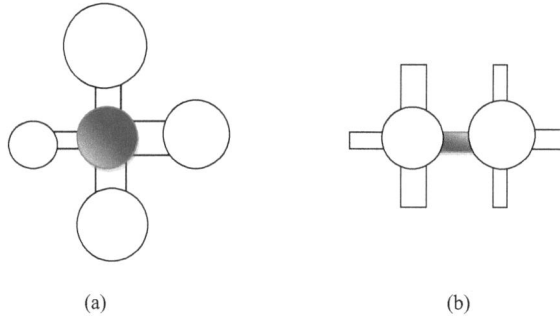

(a)   (b)

Fig. (2). Basic units for (**a**) cavities and (**b**) bonds.

The first condition is expressed as $S(R_C)$ according to Eq. (4.1). The second condition implies that the cavity cannot saturate if all concurrent bonds remain filled with gas. This last condition can be expressed as $(1 - L_{BI})^C$ where C is the connectivity of the network and L_{BI} represents the probability for a bond to be saturated and linked to the bulk of water. As pores can only exhibit one of two possible states: saturated or dry, then $1 - L_{BI} = G_{BI}$ represents the probability of the bond to be filled with gas. Accordingly, the condition that at least one bond concurrent to the considered cavity is saturated and linked to the bulk of water can be expressed as $[1 - (1 - L_{BI})^C]$. Therefore, if L_{SI} represents the probability for a cavity to saturate during a wetting process, its value is given by the product of the two conditions listed above as they must occur simultaneously:

$$L_{SI} = S(R_C)\left[1 - (1 - L_{BI})^C\right] \tag{4.3}$$

On the other hand Fig. (**2b**) shows the basic unit for bonds. This unit can be used to establish the conditions for bonds to saturate or dry. It consists of a central bond connecting two cavities each one of which is connected to *C-1* additional bonds. Based on this unit, it can be said that a bond can saturate during a wetting process only when the two following conditions are simultaneously fulfilled: a) Its size is smaller than the critical radius and therefore it can be invaded by water and b) at least one of the two sites to which it is connected is already saturated and connected to the bulk of water. The first condition can be represented by $B(R_C)$ according to Eq. (4.2). The second condition can be expressed as $[1 - (1 - L_{SI})^2]$, where $(1 - L_{SI})^2 = G_{SI}^2$ represents the probability of the two linked sites to be filled with gas at the same time. Therefore, if L_{BI} represents the probability for a bond to saturate during a wetting process, its value is given by the product of the two aforementioned conditions as they must occur simultaneously:

$$L_{BI} = B(R_C)\left[1 - (1 - L_{SI})^2\right] \tag{4.4}$$

By substituting the above relationship in Eq. (4.3), it results in

$$L_{SI} = S(R_C)\left\{1 - \left[1 - B(R_C)L_{SI}\right]^C\right\} = S(R_C)F_{SI}^s(R_C) \tag{4.5}$$

Note that the exponent *2* in Eq. (4.4) has been replaced by *1* in Eq. (4.5). This is so because L_{SI} in Eq. (4.5) represents the probability of a cavity to be filled with water. Therefore, its C converging bonds have only one additional cavity to which they are connected to and this cavity must comply with the condition of being saturated (white cavities in Fig. (**2a**)). Therefore, Eq. (4.4) transforms into $L_{BI} = B(R_C)L_{SI}$.

On the other hand, parameter $F_{SI}^s(R_C)$ represents the proportion of cavities filled with water with respect to the total number of cavities during a wetting process, in other words, it represents a saturation factor at wetting for the function $S(R_C)$ (see Fig. (**1a**)). Therefore, the volume of saturated cavities during a wetting process at the critical radius R_C ($V_{SI}^s(R_C)$) can be found by multiplying the reduction factor by the volume of cavities which sizes range from zero to R_C, in the form

$$V_{SI}^s(R_C) = F_{SI}^s(R_C)\sum_{R=0}^{R_C} V_{SR}(R)$$

where $V_{SR}(R)$ represents the volume of all sites of size R. This equation implies that there is a proportion of $F_{SI}^s(R_C)$ of the total volume of sites which size ranges from zero to R_C that saturate under wetting (see Fig. **1a**).On the other hand, by substituting Eq. (4.3) into Eq. (4.4), it results

$$L_{BI} = B(R_C)\left\{1-\left\{1-S(R_C)\left[1-(1-L_{BI})^{C-1}\right]\right\}^2\right\} = B(R_C)F_{BI}^s(R_C) \qquad \textbf{(4.6)}$$

Note that the exponent C in Eq. (4.3) has been replaced by $C-1$ in Eq. (4.6). This is so because L_{BI} in Eq. (4.6) represents the probability of a bond to be filled with water. Therefore its two connected sites have only $C-1$ additional bonds that must comply with the condition that at least one of them should be filled with water (white bonds in Fig. **2b**).

The parameter $F_{BI}^s(R_C)$ in the above relationship represents the proportion of bonds filled with water with respect to the total number of bonds which sizes range from zero to R_C, *i.e.* it represents the saturation factor at wetting for the function $B(R_C)$ (see Fig. **1a**).

In the same manner as for cavities, the volume of saturated bonds can be found with the product of the reduction factor by the volume of bonds of a certain size, in the form

$$V_{BI}^s(R_C) = F_{BI}^s(R_C)\sum_{R=0}^{R_C} V_{BR}(R)$$

where $V_{BR}(R)$ represents the volume of all bonds of size R.

It can be proved that Eqs. (4.5) and (4.6) are consistent as L_{SI} and L_{BI} are equal to zero and one when both $S(R_C)$ and $B(R_C)$ reach these same values. These equations can be solved by any iterative method and, in general, the convergence to the solution occurs in two or maximum three iterations with a tolerance of a thousandth.

In this manner, the degree of saturation at wetting (S_{wI}) for a certain value of the critical radius (R_C) is given by

$$S_{wI}(R_C) = \frac{V_{SI}^s(R_C)+V_{BI}^s(R_C)}{V_S+V_B} \qquad \textbf{(4.7)}$$

where V_S and V_B represent the total volume of sites and bonds. And because the critical radius is dependent on the value of suction according to Eq. (3.2) (Chapter

3), it is possible to plot the wetting SWRC in the axes of suction *versus* degree of saturation.

Because the probabilistic model considers a network of infinite size, the effect of the borders is not taken into account. Either way, for very large networks this effect becomes negligible. For example, if it is assumed that at the borders of the network only solids and bonds can be found, then all bonds at the borders are connected to a single site. In a cubic network made of *n* sites by side, the proportion of bonds at the border with respect to the total is approximately *3/n* which represents a very small proportion if it is acknowledged that *n* is of the order of several thousands to several millions per gram of material depending on the type of soil.

4.4. MAIN DRYING CURVE

Consider now a drying process (see Fig. **1b**). By inspecting the basic unit in Fig. (**2a**), it can be said that a cavity should comply with one of the following two conditions to remain saturated during a drying process: a) its size is smaller than the critical size and therefore it cannot be drained or b) it is larger or equal to the critical size but at the same time, all its concurrent bonds are saturated. The first condition is represented by $S(R_C)$ according to Eq. (4.1). The second condition can be expressed as $[1 - S(R_C)]L_{BD}{}^C$, where L_{BD} represents the probability for a bond to be saturated during a drying process. Then, if L_{SD} represents the probability of a cavity to remain saturated and connected to the bulk of water during a drying process, its value is obtained by the addition of the two above conditions because they are complementary

$$L_{SD} = S(R_C) + [1 - S(R_C)]L_{BD}{}^C \qquad (4.8)$$

Consider now the basic unit for bonds shown in Fig (**3b**). According to this figure, it can be established that a bond requires complying with one of the following two conditions in order to remain saturated during a drying process: a) its size is smaller than the critical radius and therefore, water cannot be displaced by gas or b) it can be invaded by gas but the two sites to which it is connected are saturated. The first condition can be written as $B(R_C)$ according to Eq. (4.2). The second condition states that the bond is able to drain (condition which is represented by $[1 - B(R_C)]$) but it remains filled with water because its two connected sites are saturated (condition represented by $L_{SD}{}^2$). Then, this second condition can be expressed as $[1 - B(R_C)]L_{SD}{}^2$. Therefore, if L_{BD} represents the probability for a bond to be saturated, its value can be represented by the addition of these two conditions as they are complementary, that is to say

$$L_{BD} = B(R_C) + [1 - B(R_C)]L_{SD}^{2} \tag{4.9}$$

And by substituting Eq. (4.9) into (4.8), it becomes:

$$L_{SD} = S(R_C) + [1 - S(R_C)]\{B(R_C) + [1 - B(R_C)]L_{SD}\}^C = S(R_C) + [1 - S(R_C)]F_{SD}^{s}(R_C) \tag{4.10}$$

Notice that the exponent *2* in Eq. (4.9) has been reduced to *1* in Eq. (4.10) because this last equation represents the probability of a cavity to be saturated and therefore, its *C* concurrent bonds are only linked to one additional cavity (blank cavities in Fig. **2a**).

Also notice that in this case the reduction factor for the term $S(R_C)$ is unity, meaning that in spite of the size of adjacent elements, all cavities whose size is smaller than the critical size remain saturated at this stage. On the other hand, those cavities larger or equal to the critical size have a proportion of $F_{SD}^{s} = \{B(R_C) + [1 - B(R_C)]L_{SD}\}^C$ saturated elements (see Fig. **1b**).

Furthermore note that the element $[1 - S(R_C)]F_{SD}$ involves the product $[1 - S(R_C)]$ $[1 - B(R_C)]$. This product is performed numerically in the model by discretizing the size distributions of cavities and bonds in the form $\{\Delta[1 - S(R_C)]\}\{\Delta[1 - B(R_C)]\}$. Because the construction principle establishes that two adjacent bonds should not intersect each other then, all products where the size of bonds divided by a factor $\sqrt{2}$ (as it is considered that the two adjacent bonds have the same size) is larger than the size of cavities, are discarded. This consideration has to be applied to all factors involving the product between the size distribution of cavities and bonds.

On the other hand, by substituting Eq. (4.8) in (4.9) it results in

$$L_{BD} = B(R_C) + [1 - B(R_C)]\{[1 - S(R_C)]L_{BD}^{C-1}\}^2 = B(R_C) + [1 - B(R_C)]F_{BD}^{s}(R_C) \tag{4.11}$$

Notice that the exponent *C* in Eq. (4.8) has been reduced to *C-1* in Eq. (4.11) because this last equation represents the probability of a bond to be saturated and therefore, the sites connected to this bond only have *C-1* additional bonds that require to be liquid filled (bonds in white in Fig. (**2b**). Also note that the product $[1 - B(R_C)]S(R_C)$ has been removed from Eq. (4.11). This is so because it involves the product of bonds of larger size $(1 - B(R_C))$ than the size of sites $S(R_C)$, as it can be verified in Fig. (**1b**). And, because the construction principle states that sites must be larger than the bonds they are connected to, this product becomes null. Finally, it can be verified that Eqs. (4.10) and (4.14) are consistent because L_{SD} as well as L_{BD} become zero and unity when both $B(R_C)$ and $S(R_C)$ reach these same values.

With the above equations it is possible to determine the volume of bonds and sites that remain saturated during a drying process when the critical radius reaches the value R_C, as shown below.

These equations indicate that during a drying process all pores smaller than the critical size remain saturated as suction cannot dry them whereas those bonds and sites larger or equal to the critical size have a proportion of F_{BD}^s and F_{SD}^s of saturated elements, respectively (see Fig. (**1b**)).

Accordingly, the degree of saturation of a sample subject to drying ($S_{wD}(R_C)$) is given by the relationship

$$S_{wD}(R_C) = \frac{V_{SD}^s(R_C) + V_{BD}^s(R_C)}{V_S + V_B} \tag{4.12}$$

And because the value of R_C depends on suction (Eq. (3.2), Chapter 3), then it is possible to plot the drying SWRC in the axes of degree of saturation *versus* suction.

The degree of saturation given by Eqs. (4.12) and (4.7) varies from 1 to 0, however real soils subject to wetting-drying cycles never reach these values. Because of this, these equations are affected by two reduction factors: one representing the maximum and the other the residual degree of saturation of the soil as explained before.

It can be verified that these equations obtained by means of the basic unit concept are the same as those derived from a purely analytical procedure [59], although in this case they have been obtained in a simpler and more rational way. In addition, this procedure allows introducing the solid phase into the model as it is shown below.

4.5. SATURATED AND DRY VOLUMES

Once the volumes of cavities and bonds filled with water have been defined, it is possible to determine the saturated and the dry volumes during a wetting or drying process. For this purpose it is convenient to consider the basic solid unit depicted in Fig. (**3**). It represents the 2D case where a solid is encircled by four cavities and four bonds. These elements are called the surrounding pores and are shown in gray tones. In addition to these elements there are the external elements (shown in white) consisting in eight bonds and eight cavities. To these elements follow the farther external sites and bonds (not shown in the figure). Table **1** shows the number of surrounding, external and farther external sites and bonds in a bi- and tri-dimensional network. It also shows the equations defining the number of elements for each case. These equations are used to determine the probability of surrounding pores to be saturated or dry.

Consider that a soil sample undergoes a wetting process. Initially all pores are dry and suction is very large. Then, suction reduces by steps and the smallest pores are the first to saturate. By inspecting Fig. (**3**) it can be established that all pores surrounding a solid saturate when the following two conditions are fulfilled: a) all bonds and cavities surrounding the solid are smaller than the critical size, *i.e.* they can be intruded by water and, b) simultaneously at least one external bond is already saturated and connected to the bulk of water.

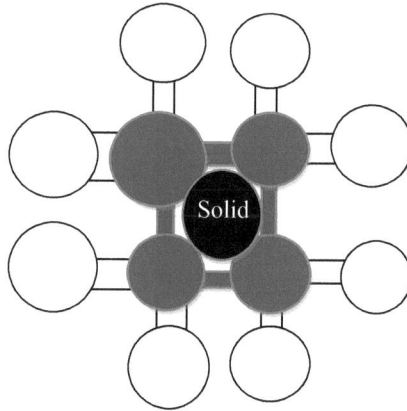

Fig. (3). Basic unit for solids.

Table 1. Number of surrounding, external and farther external elements in a basic solid unit.

Type	Element	2D	3D	Equation
Surrounding	Cavities	4	8	2(C-2)
	Bonds	4	12	4(C-3)
External	Cavities	8	24	C(C-2)
	Bonds	8	24	C(C-2)
Farther external	Cavities	12	72	6(C-2)(C-3)
	Bonds	20	96	C(C-2)²+2(6-C)

The first condition can be written as $[S(R_C)]^{2(C-2)}[B(R_C)]^{4(C-3)}$ where the exponents *2(C-2)* and *4(C-3)* represent the number of surrounding cavities and bonds, respectively (refer to Table **1**). The second condition can be expressed as $1 - [1 -$

$L_{BI1}]^{C(C-2)}$ where the exponent $C(C-2)$ represents the number of external bonds (see Table **1**) and L_{BI1} represents the probability of an external bond to be liquid filled. When all the surrounding pores of a solid are saturated, they form a saturated unit. Therefore, the probability of having a saturated unit at wetting ($L_{\Im I}$) results from the product of the two aforementioned conditions (as they must be fulfilled simultaneously), in the form

$$L_{\Im I} = \left[S(R_C)\right]^{2(C-2)}\left[B(R_C)\right]^{4(C-3)}\left\{1 - \left[1 - L_{BI_1}\right]^{C(C-2)}\right\} \tag{4.13}$$

Parameter L_{BI1} can be obtained from Eqs. (4.3) and (4.4) but with different exponents as the number of connected elements is different for this case. For example, exponent C in Eq. (4.3), representing the number of bonds connected to a single cavity in the basic cavity unit, transforms into $\dfrac{C(C-2)^2 + 2(6-C)}{C(C-2)}$ which represents the number of farther external bonds connected to a single external cavity. This value is obtained by dividing the corresponding equations indicated in Table **1**. Also, exponent 2 in Eq. (4.6) transforms into 1 as all external bonds are connected to a single external site (refer to Table **1**). Therefore, the expression for L_{BI_1} results in

$$L_{BI_1} = B(R_C)S(R_C)\left[1 - \left[1 - L_{BI_1}\right]^{\frac{C(C-2)^2 + 2(6-C)}{C(C-2)}}\right]$$

On the other hand and as has been established, a solid appertains to the dry fraction when all its surrounding pores are filled with gas. Consider again the basic unit shown in Fig. (3). By inspecting this figure it can be concluded that all pores (cavities and bonds) surrounding a solid remain dry as long as all surrounding cavities keep dry. Therefore, if $G_{\Im I}$ represents the probability of a solid to be surrounded by pores filled with gas and, G_{SI} is the probability of a cavity to be filled with gas, the above condition writes

$$G_{\Im I} = G_{SI}^{2(C-2)} = \left(1 - L_{SI_1}\right)^{2(C-2)} \tag{4.14}$$

where the exponent $2(C-2)$ represents the number of surrounding cavities and L_{BI_1} represents the probability of a surrounding site to be liquid filled. The value of parameter L_{BI_1} can be obtained from a relationship similar to Eq. (4.5) except that exponent C in this last equation is substituted by $C/2$ which represents the number of external bonds connected to a single surrounding site, according to Table **1**. By doing this substitution, Eq. (4.5) transforms into

$$L_{SI_1} = S(R_C)\left\{1 - \left[1 - B(R_C)L_{SI_1}\right]^{C/2}\right\}$$

By solving this equation, the value of $G_{\tilde{s}I}$ can be obtained from Eq. (4.14).

Consider now a drying process. Initially all pores are saturated and suction is equal to zero. Then, suction increases by steps and the largest pores will be the first to dry. By inspecting Fig. (**3**) it can be concluded that all surrounding bonds remain saturated as long as all surrounding sites also do so. Therefore, the probability that all pores surrounding a solid are saturated during a drying process ($L_{\tilde{s}D}$) is equal to the probability that all surrounding sites remain saturated

$$L_{\tilde{s}D} = L_{SD_1}^{2(c-2)} \tag{4.15}$$

where the exponent is the number of sites surrounding a solid and L_{SD_1} represents the probability of surrounding sites to be liquid filled during a drying process. The relationship to determine L_{SD_1} is similar to Eq. (4.10) except that exponent C transforms into C/2 which represents the number of external bonds connected to a single surrounding site according to Table **1**. Therefore, by doing this substitution, Eq. (4.10) transforms into

$$L_{SD_1} = S(R_C) + \left[1 - S(R_C)\right]\left\{B(R_C) + \left[1 - B(R_C)\right]L_{SD_1}\right\}^{C/2} \tag{4.16}$$

By solving Eq. (4.16) it is possible to obtain the value of L_{SD_1} from Eq. (4.15).

On the other hand, the conditions required for all surrounding pores of a solid to dry during a drying process are the following: a) all surrounding cavities and bonds must be able to dry and b) at least one external bond should already be dry and connected to the bulk of gas. The first condition is written as $[1 - S(R_C)]^{2(C-2)}[1 - B(R_C)]^{(4-3)}$. The second condition can be expressed as $1 - (1 - G_{BD_1})^{C(C-2)} = 1 - L_{BD_1}^{C(C-2)}$ where G_{BD_1} and L_{BD_1} represent the probability of an external bond to be filled with gas and liquid, respectively. Then, if $G_{\tilde{s}D}$ represents the probability of finding dry solid units during a drying process, its value is given by the product of the two aforementioned conditions as they must occur simultaneously

$$G_{\tilde{s}D} = \left[1 - S(R_C)\right]^{2(C-2)}\left[1 - B(R_C)\right]^{4(C-3)}\left(1 - L_{BD_1}^{C(C-2)}\right) \tag{4.17}$$

The value of L_{BD_1} can be obtained from the combination of Eqs. (4.8) and (4.9) but with different exponents according to the number of connected elements.

According to Table **1**, exponent C in Eq. (4.8) transforms into $\dfrac{C(C-2)^2 + 2(6-C)}{C(C-2)}$

which represent the number of farther external bonds connected to a single external cavity while exponent 2 in Eq. (4.9) transforms into 1 because it represents the number of external bonds connected to a single external cavity (refer to Table 1). By combining the resulting equations, it becomes

$$L_{BD_1} = B(R_C) + [1 - B(R_C)]\left\{[1 - S(R_C)]L_{BD_1}^{\frac{C(C-2)^2+2(6-C)}{C(C-2)}}\right\}$$

By solving the relationship above, the value of $G_{\Im D}$ can be obtained from Eq. (4.17). Observe that the product $S(R_C)[1 - B(R_C)]$ becomes null as discussed before. With these relationships it is now possible to define the distribution of saturated cavities and bonds as a function of their size. For example, in the case of sites, Eq. (4.13) can be rewritten in the following form

$$L_{\Im I} = S(R_C)\,S(R_C)^{2C-5}\,B(R_C)^{4(C-3)}\left\{1 - [1 - L_{BI_1}]^{C(C-2)}\right\} = F_{SI}^s\,S(R_C) = F_{SI}^s \int_0^{R_C} V_{RS}(R)dR$$

where F_{SI}^s represents the saturation factor for cavities in a solid unit. Then, the differential volume of saturated cavities of size R is represented as

$$dV_S^s = F_{SI}^s\,V_S\,V_{RS}(R)dR$$

And the total volume of saturated cavities (V_S^s) is:

$$V_S^s = F_{SI}^s\,V_S \int_0^{R_C} V_{RS}(R)dR = F_{SI}^s\,S(R_C)V_S = L_{\Im I}\,V_S$$

In the same way the distribution of the volume of saturated bonds V_B^s as a function of their size can be obtained. In that case, the saturation factor for the relative volume of bonds (F_{BI}^s) results from Eq. (4.13), rewritten in the following form

$$L_{\Im I} = B(R_C)\,B(R_C)^{4C-13}\,S(R_C)^{2(C-2)}\left\{1 - [1 - L_{BI_1}]^{C(C-2)}\right\} = F_{BI}^s\,B(R_C)$$

Then, the differential volume of saturated bonds of size R is given by $dV_B^s = F_{BI}^s\,V_B V_{RB}(R)dR$ and, the total volume of saturated bonds (V_B^s) is

$$V_B^s = F_{BI}^s\,V_B \int_0^{R_C} V_{RB}(R)dR = F_{BI}^s\,B(R_C)V_B = L_{\Im I}\,V_B$$

Finally, because there is a correlation in sizes between cavities and solids as explained before, the volume of saturated solids ($V_{\Im I}^s$) can be determined as

$V^s_{\tilde{s}I} = L_{\tilde{s}I} V_{\tilde{s}}$ Once the values of saturated cavities, bonds and solids have been defined, it is possible to define the volume of the saturated fraction

$$f^s = \frac{V^s}{V} = \frac{V^s_S + V^s_B + V^s_{\tilde{s}}}{V}$$

The same procedure can be applied to define the dry fraction. For example, for drying process, Eq. (4.17) can be rewritten as

$$G_{\tilde{s}D} = \left(1 - L_{BD_1}{}^{C(C-2)}\right)\left[1 - B(R_C)\right]^{4(C-3)}\left[1 - S(R_C)\right]^{2C-5}\left[1 - S(R_C)\right] = F^d_{SD}\left[1 - S(R_C)\right]$$

where F^d_{SD} represents the dry factor for cavities in the solid unit during a drying process. Then, the differential volume of dry cavities of a certain size $(dV^d_S(R))$ can be found by multiplying the dry factor F^d_{SD} by the volume of cavities of that specific size. Finally, the total volume of saturated cavities is obtained by the addition of all saturated cavities from size R_C to the maximum size R_{max}, that is to

say $V^d_S = \int_{R_C}^{R_{max}} dV^d_S(R) = F^d_{SD} V_S \int_{R_C}^{R_{max}} V_{RS}(R)$. By applying the same procedure, the volume

of dry bonds V^d_B can be obtained. Moreover, the volume of solids of the dry fraction is given by $V^d_{\tilde{s}} = G_{\tilde{s}D} V_{\tilde{s}}$ and finally, the dry fraction is

$$f^d = \frac{V^d}{V} = \frac{V^d_S + V^d_B + V^d_{\tilde{s}}}{V}$$

Once f^s and f^d have been established it is possible to define the value of the unsaturated fraction f^u using Eq. (2.12), Chapter 2. Furthermore, the degree of saturation of the unsaturated fraction can be obtained by dividing the volume of water by the volume of voids both belonging to the unsaturated fraction. The volume of water of the unsaturated fraction is obtained by subtracting the volume of bonds and cavities belonging to the saturated fraction from the volume of pores filled with water. On the other hand, the volume of voids of the unsaturated fraction can be obtained by subtracting the saturated and dry volumes of bonds and cavities from the total volume of voids, that is to say

$$S^u_w = \frac{V^u_w}{V^u_v} = \frac{V_w - V^s_w}{V_v - V^s_v - V^d_v} = \frac{V_w - V^s_S - V^s_B}{V_v - V^s_S - V^s_B - V^d_S - V^d_B}$$

Once parameters f^s, f^u and S^u_w have been established, it is possible to determine the mean effective stress (Eq. (2.14), Chapter 2) and the shear strength (Eq. (2.18), Chapter 2) of a soil subject to any value of suction during a wetting or drying path.

In principle, the parameters required by the porous-solid model are the pore (PSD) and the grain size distribution (GSD), the voids ratio, the connectivity and a shape parameter for the fine fraction. All these parameters are discussed below.

When the PSD is obtained from porosimetry tests, it is possible to discriminate the volume of mesopores from such of macropores because usually porosimetry tests performed on soils show a bimodal distribution: the one with smaller sizes corresponds to the mesopores and the other with larger sizes corresponds to the macropores [45]. However, these tests do not permit the determination of the size distribution of bonds mainly because the volume of these elements is negligible compared to that of cavities. Therefore, in order to define the size distribution for bonds, use can be made of two properties of SWRCs: the first one establishes that there is a unique relationship between the PSD and the SWRCs [60]. The second is that the drying branch depends mainly on the size distribution of bonds while the wetting branch is dependent mainly on the size distribution of cavities [28]. In this way, by fitting the numerical drying curve with the experimental results, it is possible to define the size distribution for bonds. This process begins by proposing a size distribution for these elements while the size distribution for cavities is obtained from a porosimetry test. Then the numerical drying SWRC is compared with the experimental one and the size distribution for bonds is subsequently adjusted until the best fit is obtained. If porosimetry data is not available, then both SWRCs are required and the adjusting process includes the size distributions for macropores, mesopores and bonds. The PSD and GSD of the material can be introduced directly into the model in the form of tables or they can be adjusted using proper mathematical functions. Presently, the model uses single, double or triple logarithmic normal distributions to adjust the experimental data. These functions have proven to be sufficiently flexible in order to accurately simulate the pore and GSDs for different types of soils. Direct porosimetry tests such as image analysis of micrographs or indirect methods such as mercury intrusion porosimetry (MIP) or nitrogen adsorption can be used to determine the PSD of a soil.

The connectivity of a 2D and a 3D network is 4 and 6, respectively. However, direct determination techniques indicate that real soils show connectivity values between 2 and 6 as mention before [55].

Finally, as the size distribution of the solid particles plays an important role in the determination of the volume of saturated solids, it is important to take into account the hypothesis considered to obtain this distribution. For example, one of the most widely used methods to obtain the GSD of fine soils is the hydrometer test. However, this method assumes that solid particles have rounded shapes. This hypothesis may be adequate for the grains of sands but not for fine particles and especially not for clays which in general show flat shapes. For this reason, a shape factor (S_f) needs to be introduced into the porous-solid model, as explained before. This parameter adjusts the real volume of fine material ensuring that the numerical voids ratio corresponds to the experimental one. The shape factor comes close to unity for rounded sands but reduces drastically in the case of plastic clays.

4.6. SCANNING CURVES

When an inversion on a wetting or drying process takes place, the resulting paths are called scanning curves. The mechanisms of wetting or drying for sites and bonds in this case are exactly the same as those described for the boundary curves, except that the initial conditions change, as it is explained below.

4.6.1. Drying –Wetting Cycle

Fig. (**1b**) shows the proportion of pores filled with gas at a certain stage of a drying process. In the case of a drying-wetting cycle, only those bonds and sites already invaded by gas during the drying stage (G_{BDR} and G_{SDR}, respectively) can be replenished with water, and therefore the values of L_{SI} and L_{BI} in Eqs. (4.5) and (4.6) should be substituted by the terms (L_{SDI} - L_{SDR})/G_{SDR} and ($L_{BDI} - L_{BDR}$)/G_{BDR}, where L_{SDI} and L_{BDI} represent the probability for a site or a bond, respectively, to be liquid-filled after an inversion in drying and L_{SDR} and L_{BDR} represent the probability for a site and a bond to be liquid-filled at the moment of inversion, respectively. Additionally, the probability functions $S(R_C)$ and $B(R_C)$, should also be scaled according to the following equations

$$S_A(R_C) = \frac{S(R_C) - S_R(R_C)}{1 - S_R(R_C)}, \qquad B_A(R_C) = \frac{B(R_C) - B_R(R_C)}{1 - B_R(R_C)}$$

where $S_A(R_C)$ and $B_A(R_C)$ are the adjusted probability function for sites and bonds, respectively. Finally, knowing that $G_{BDR} = 1 - L_{BDR}$ and $G_{SDR} = 1 - L_{SDR}$ Eqs. 4.5 and 4.6 transform into

$$L_{SDI} = (1 - L_{SDR}) S_A(R_c) \left[1 - \left(1 - \frac{L_{SDI} - L_{SDR}}{1 - L_{SDR}} B_A(R_C) \right)^C \right] + L_{SDR}$$

$$= (1 - L_{SDR}) S_A(R_c) F_{SDI}^s(R_C) + L_{SDR}$$

(4.18)

$$L_{BDI} = (1 - L_{BDR})B_A(R_C)\left\{1 - \left\{1 - S_A(R_C)\left[1 - \left(1 - \frac{L_{BDI} - L_{BDR}}{1 - L_{BDR}}\right)^{C-1}\right]\right\}^2\right\} + L_{BDR}$$

(4.19)

$$= (1 - L_{BDR})B_A(R_C)F_{BDI}^s(R_C) + L_{BDR}$$

The above equations ensure that the adjusted probability functions for sites and bonds ($S_A(R_C)$ and $B_A(R_C)$ respectively) continue varying from one to zero and thus the equations describing L_{BDI} and L_{SDI} remain auto-consistent.

With the above equations it is possible to determine the volume of bonds and sites that remain saturated once the inversion in the wetting process initiates

$$V_{SDI}^s(R_C) = \sum_{R=0}^{R_R} V_{SR}(R) + \frac{1 - L_{SDR}}{1 - S(R_R)}\left[F_{SDI}^s(R_C)\sum_{R=R_R}^{\infty} V_{SR}(R)\right]$$

$$V_{BDI}^s(R_C) = \sum_{R=0}^{R_R} V_{BR}(R) + \frac{1 - L_{BDR}}{1 - B(R_R)}\left[F_{BDI}^s(R_C)\sum_{R=R_R}^{\infty} V_{BR}(R)\right]$$

where R_R and R_C represent the critical radius at the moment of inversion and the critical radius during the wetting stage, respectively. Parameters $F_{SDI}^s(R_C)$ and $F_{BDI}^s(R_C)$ can be obtained from Eqs. 4.18 and 4.19 in the form

$$F_{SDI}^s(R_C) = \left[1 - \left(1 - \frac{L_{SDI} - L_{SDR}}{G_{SDR}}B_A(R_C)\right)^C\right]$$

$$F_{BDI}^s(R_C) = \left\{1 - \left\{1 - S_A(R_C)\left[1 - \left(1 - \frac{L_{BDI} - L_{BDR}}{G_{BDR}}\right)^{C-1}\right]\right\}^2\right\}$$

The degree of saturation of the system is obtained as expressed in Eq. (4.7), except that $V_{SI}^s(R_c)$ and $V_{BI}^s(R_c)$ substitute for $V_{SDI}^s(R_c)$ and $V_{BDI}^s(R_c)$, respectively.

4.6.2. Wetting-Drying Cycle

When an inversion occurs at a certain stage of a wetting path, the distribution of sites and bonds invaded by water is, approximately, that shown in Fig. (**1a**).

Therefore, during the drying stage, only those sites and bonds already invaded by fluid during the wetting phase are able to drain. Consequently, the limits of the probability for a site or a bond to dry during the drying process goes from their value at the moment of inversion (L_{SIR} and L_{BIR}, respectively) to zero for a completely dry soil. Therefore, the values of L_{BD} and L_{SD} in Eqs. 4.9 and 4.10, should be scaled in the form of L_{BID}/L_{BIR} and L_{SID}/L_{SIR}, respectively, where L_{BID} and L_{SID} represent the probability for a bond and a site to be liquid-filled at drying after being subjected to wetting, respectively. The probability functions $S(R_C)$ and $B(R_C)$, should also be scaled according to the following equations

$$S_A(R_C) = \frac{S(R_C)}{S_R(R_C)}, \qquad\qquad B_A(R_C) = \frac{B(R_C)}{B_R(R_C)}$$

where $S_A(R_C)$ and $B_A(R_C)$ represent the adjusted values of the probability functions for sites ($S(R_C)$) and bonds ($B(R_C)$), while $S_R(R_C)$ and $B_R(R_C)$ represent the probability values for sites and bonds at the moment of inversion, respectively.

With these considerations, Eqs. 4.10 and 4.11 transform into

$$L_{SID} = \left\{ S_A(R_C) + [1 - S_A(R_C)]\{B_A(R_C) + [1 - B_A(R_C)](L_{SID}/L_{SIR})\}^C \right\} L_{SIR}$$

$$= \left\{ S_A(R_C) + [1 - S_A(R_C)]F_{SID}^s \right\} L_{SIR}$$

$$\text{(4.20)}$$

$$L_{BID} = \left\{ B_A(R_C) + (1 - B_A(R_C))\{1 - S_A(R_C)(L_{BID}/L_{BIR})^{C-1}\}^2 \right\} L_{BIR}$$

$$= \left\{ B_A(R_C) + (1 - B_A(R_C))F_{BID}^s \right\} L_{BIR}$$

$$\text{(4.21)}$$

The above equations ensure that the adjusted probability functions for sites ($S_A(R_C)$) and bonds ($B_A(R_C)$) remain varying from one to zero, and thus the equations defining L_{BID} and L_{SID} remain auto-consistent. In this form, when an inversion takes place, the sign of the increment of the probability functions $S(R_C)$ or $B(R_C)$ change, and they continue increasing or reducing until a new inversion takes place.

With the above equations it is possible to determine the volume of bonds and sites that remain saturated once the wetting of the sample stops and the drying process initiates

$$V_{SID}^s(R_C) = \frac{L_{SIR}}{S(R_R)} \left[\sum_{R=0}^{R_{CR}} V_{SR}(R) + F_{SID}^s(R_C) \sum_{R=R_{CR}}^{R_R} V_{SR}(R) \right]$$

$$V_{BID}^s(R_C) = \frac{L_{BIR}}{B(R_R)} \left[\sum_{R=0}^{R_{CR}} V_{BR}(R) + F_{BID}^s(R_C) \sum_{R=R_{CR}}^{R_R} V_{BR}(R) \right]$$

where R_R and R_C represent the critical radius at the moment of inversion and the critical radius during the drying stage, respectively. Parameters $F_{SDI}^s(R_C)$ and $F_{BDI}^s(R_C)$ can be obtained from Eqs. (4.20) and (4.21) in the form

$$F_{SID}^s(R_C) = \left\{ B_A(R_C) + \left[1 - B_A(R_C) \right] (L_{SID}/L_{SIR}) \right\}^C$$

$$F_{BID}^s(R_C) = \{[1 - S_A(R_C)](L_{BID}/L_{BIR})^{C-1}\}^2$$

The degree of saturation of the system is obtained as expressed in Eq. (4.12), except that $V_{SD}^s(R_C)$ and $V_{BD}^s(R_C)$ substitute for $V_{SID}^s(R_C)$ and $V_{BID}^s(R_C)$, respectively.

The above equations can be used to simulate the secondary boundary curves as well as the scanning curves when they are plotted either on the axes of equivalent volumetric water content or degree of saturation *versus* suction. The equivalent volumetric water content Θ, was defined by van Genuchten [61] in the form $\Theta = (\theta{-}\theta_r)/(\theta_s{-}\theta_r)$, where θ_r and θ_s represent the residual and saturated volumetric water content, respectively, and θ is the current volumetric water content of the sample.

The model presented above does not consider the volumetric strains when soils are subject to loading or wetting-drying cycles [45]. However, in Chapter 11, the influence of volumetric strains on the PSD is analyzed and included in the porous-solid model to generate fully coupled constitutive models.

With the above equations, a very simple computer program can be created which generates results within seconds. This model offers important advantages when compared with computational network models. Usually these last models require heavy programs that are difficult to manipulate and require several hours to produce results for large networks. Another advantage of the probabilistic model is that all pore and solid sizes are well-represented, no matter if their size distribution curves involve several orders of magnitude, which is certainly not the case for computational network models. This has important consequences regarding the influence of the size of the network on the results especially for materials with large size distributions. While the probabilistic model considers an infinite network, computational models require defining the size of the network and it is possible that the largest network that a PC can handle may not be sufficient to nullify the size effect. Additionally, the computer code of the probabilistic porous-solid model can be easily adapted to other computer programs to develop general constitutive models for soils.

4.7. ASSESMENT OF THE PROBABILISTIC MODEL

The probabilistic model can be evaluated by comparing its results with those produced by a computational network model using a simple normal distribution with low standard deviation for the sizes of sites and bonds. Fig. (**4**) shows this comparison for the following data $\bar{R}_S = 1.0$ µm, $\delta_S = 2.5$, $\bar{R}_B = 0.2$ µm and $\delta_B = 2.0$, where $\bar{R}_S$ and $\bar{R}_B$ represent the medium size for sites an bonds whereas δ_S and δ_B are the standard deviation for sites and bonds, respectively. The connectivity in the probabilistic model was taken as four (2D case) and the size of the network in the computational network model was 500 x 500. Notice that the results produced by both models regarding the SWRC are practically the same. The most important difference is that the drying curve generated by the probabilistic model is slightly displaced to the left as compared to that of the computational model. Also, the wetting curve of the probabilistic model shows a more gradual variation close to saturation than that of the computational model. This last detail is linked to the small number of pores of the largest size that result during the determination of the number of pores of each size in the computational network model.

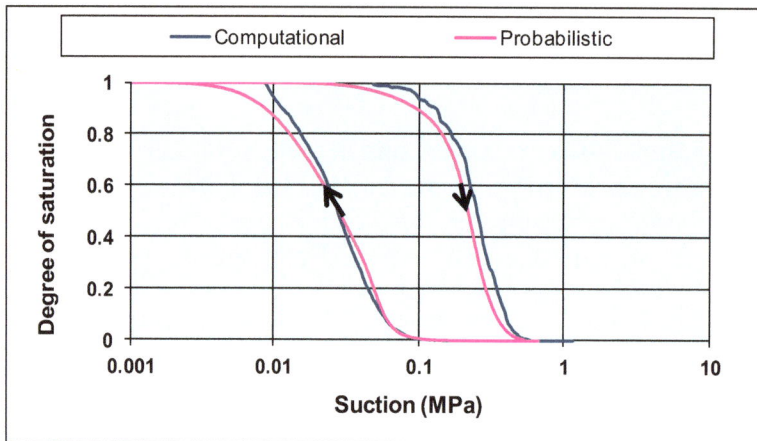

Fig. (4). Results comparison between the computational and probabilistic model.

To validate the porous-solid model described above, the experimental results reported by Brown [62] and Enustun and Enuysal [63] are used. Brown [62] reported the main hysteresis loop and some scanning curves for a sample of Vycor glass using gas xenon in isothermal conditions. His experimental results are reported in Fig. (**5**) in the axes of equivalent volumetric water content *versus* suction. Additionally, Enustun and Enuysal [63] determined the size distribution of this same material by filling the pores with metal, leaching away the glass and imaging the residue in an electron microscope. Their results are presented in Fig. (**6**) in the axes of frequency *vs*. pore radius.

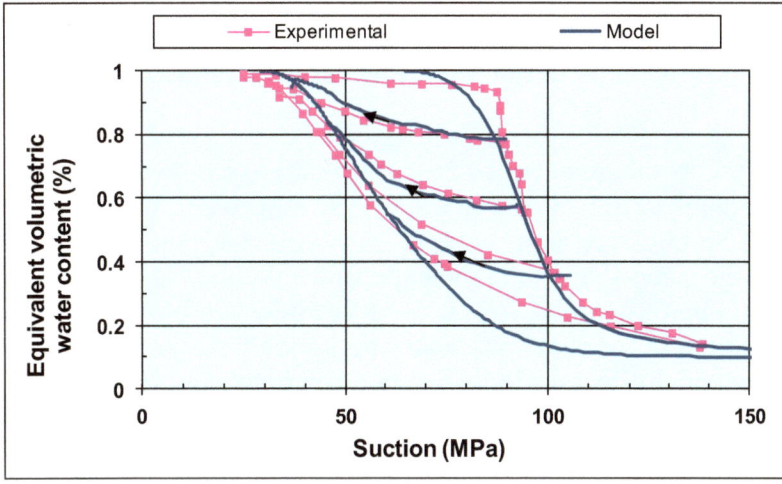

(a)

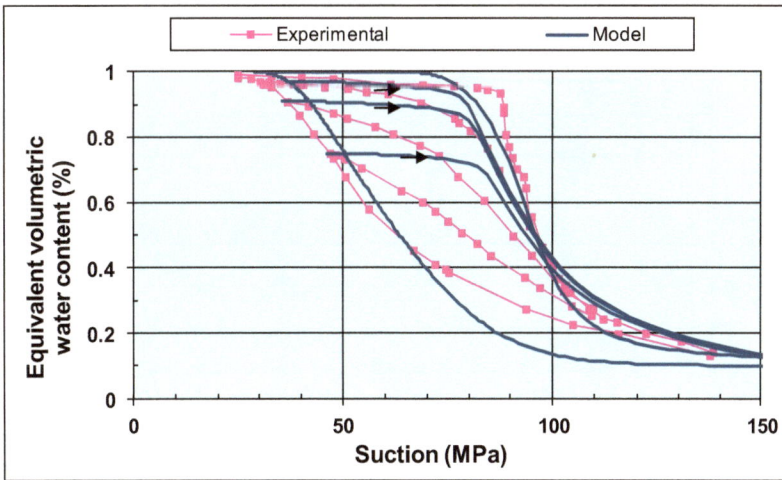

Fig. (5). Numerical and experimental main retention curves and scanning curves for (**a**) drying-wetting and (**b**) wetting-drying cycles. Experimental results from [62].

Fig. (**5**) also shows the comparison between numerical and experimental results for the main wetting and drying curves as well as for various scanning curves at wetting and drying. In this case, the main numerical adsorption and desorption curves were fitted with the experimental results by means of the iterative PSD method earlier described. For this case a normal distribution for mesopores and bonds was sufficient to properly describe the wetting and drying SWRC. Given the uniform porosity of the Vycor glass, the use of macropores to simulate the SWRC was not necessary. The size distribution parameters used by the porous-solid model to generate these figures were: $\overline{R}_S$=0.0020μm, δ_S = 0.001μm, $\overline{R}_B$=0.0013 μm, δ_B = 0.0003 μm. The numerical scanning curves were obtained by defining the water content at which the inversion from drying to wetting takes place. A good general agreement between experimental and numerical results can be observed.

Fig. (**6**) also presents the numerical size distribution of cavities obtained from the iterative procedure to obtain the main drying and wetting curves shown in Fig. (**5**). It can be observed that the numerical size distribution for sites shows approximately the same shape than the experimental data. However, a horizontal displacement of approximately 1 μm is observed between both curves. A similar result was obtained by Mason [64] when he applied the percolation theory to reproduce the main hysteresis loops of the Vycor glass. This difference was explained by Mason [64] as the result of the reduction in the thickness of the adsorbed water layer when casting the metal into the pores. In any case, this difference is rather small and shows the degree of precision that the model can reach when simulating wetting-drying processes.

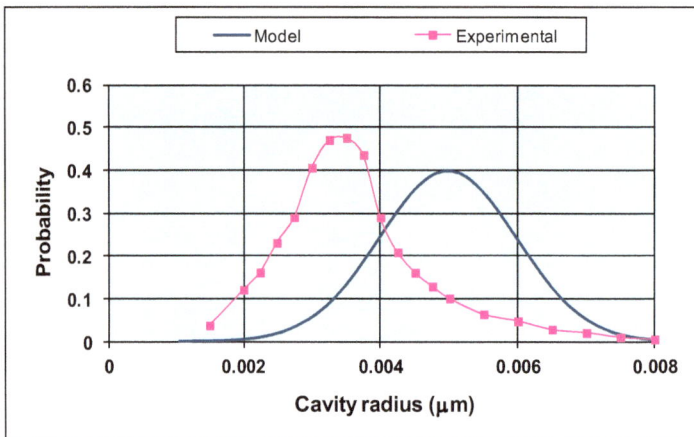

Fig. (6). Numerical and experimental size distribution for cavities (experimental results from [63]).

4.8. PARAMETRIC ANALYSIS

Figs. (**7-9**) have been prepared in order to observe the influence of the porosimetry on the wetting and drying SWRC. For simplicity, a logarithmic normal distribution for bonds and sites has been used. In a logarithmic normal distribution the only parameters required are the mean size ($\bar{R}$) and the standard deviation (δ) for each element. In addition, it is necessary to define the length of the bonds.

The wetting and drying curves in Fig. (**7**) were obtained using the following parameters: $\bar{R}_S$ = .03 μm, R_B = .003 μm, δ_S = 3.5 and δ_B = 3.5. For Fig. (**8**) the following parameters were used: $\bar{R}_S$ =.03 μm, R_B =.003 μm, δ_S = 2.5 and δ_B = 2.5. Finally, Fig. (**9**) was obtained using $\bar{R}_S$ =.1 μm. $\bar{R}_S$ =.03 μm, δ_S = 3.5 and δ_B = 3.5. The considered length of the bonds was 0.001 μm.

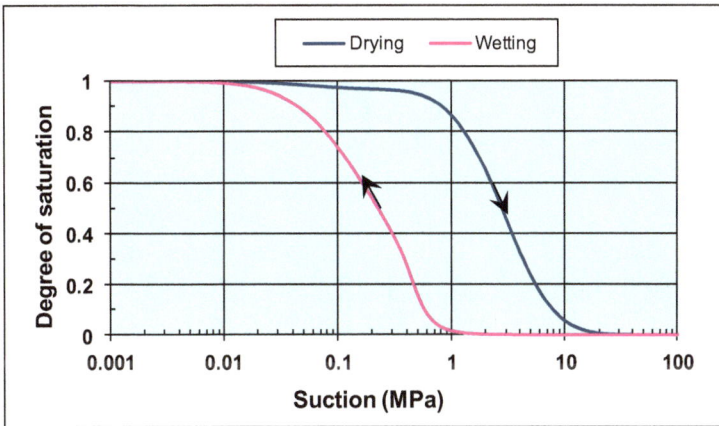

Fig. (7). Model results with $\bar{R}_S$ =.03 μm, $\bar{R}_B$ =.003 μm, δ_S =3.5 and δ_B =3.5.

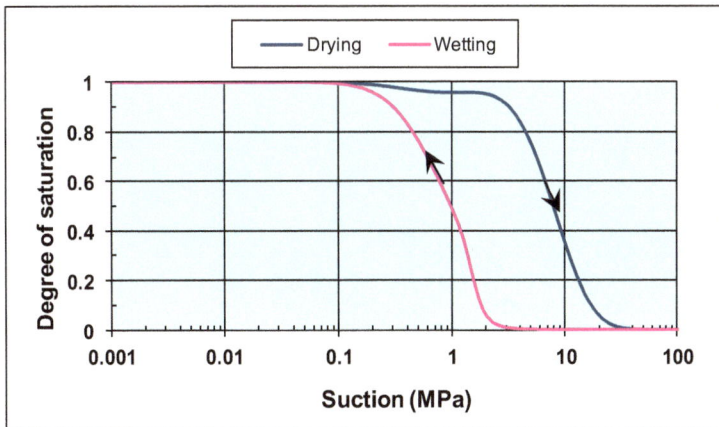

Fig. (8).   Model results with $\bar{R}_S$ =.03 μm, $\bar{R}_B$ =.003 μm, δ_S =2.5 and δ_B =2.5.

When these figures are compared, the influence of each parameter can be determined. For example, when comparing Figs. (**7** and **8**), the effect of the standard deviation on bonds and sites can be observed. When comparing Figs. (**7** and **9**), the effect of the mean size on bonds and sites is observed and finally when comparing Figs. (**8** and **9**), the combined effect of mean size and standard deviation is noticed. According to these comparisons, it can be concluded that the standard deviation of mesopores (δ_S) and bonds (δ_B) defines the extension of the curves on the axis of suction. This is so because this parameter defines the range of values of the pores for each element. Then, large standard deviations represent well graded materials with many different sizes of pores and the SWRC extends along the suction axis. In contrast, small standard deviations represent poorly graded materials with uniform pore sizes and therefore, the SWRC appears more vertical.

On the other hand, the mean size of sites ($\bar{R}_S$) and bonds ($\bar{R}_B$) defines the position of the wetting and drying curves in the axis of suction, respectively. For example, small mean sizes indicate the presence of fine soils and the curves are located at the zone of large suctions. On the contrary, large mean values indicate the presence of granular soils and the curves appear in the zone of small suctions. Also, as pointed out before, the parameters for sites affect mainly the wetting curve while those for bonds affect the drying curve. Finally, when the mean values of sites and bonds approach each other, the curves also get close together and vice versa. The length of bonds modifies the volume occupied by these elements but has no major influence on these curves.

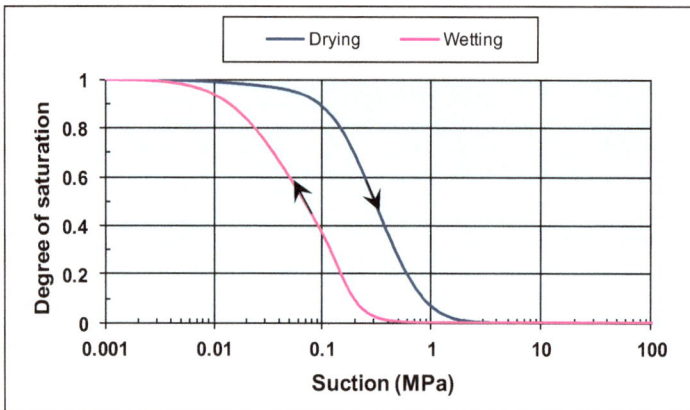

Fig. (9). Model results with R_S =.1 μm, R_B =.03 μm, δ_S =3.5 and δ_B =3.5.

CHAPTER 5

Applications of the Porous-Solid Model

Abstract: In the previous chapter, a probabilistic porous-solid model with the ability to simulate both branches of the soil-water retention curve was developed. In this chapter, the model is used to interpret more realistically the results of mercury intrusion porosimetry tests. Moreover, it is used to obtain the pore size distribution of soils employing both branches of the soil-water retention curve as data. The numerical and experimental comparisons for different soils show that the model approximately reproduces the pore size distribution obtained from mercury intrusion porosimetry tests. Finally, a procedure to fit the numerical with the experimental soil-water retention curves in order to obtain the pore size distribution of soils is presented.

Keywords: Contact angle, Critical radius, Grain size distribution, Hydro-mechanical coupling, Logarithmic normal distribution, Macropores, Mean size, Mercury intrusion porosimetry tests, Mesopores, Micropores, Pore size distribution, Relative volume, Scanning electron micrographs, Soil mixtures, Soil-water retention curve, Standard deviation, Superficial tension.

5.1. INTRODUCTION

One of the most popular methods to obtain the pore size distribution (PSD) of soils is the mercury intrusion porosimetry (MIP) tests. MIP tests are made in pressure chambers filled with mercury (which is a non-wetting fluid), where a moisture-free soil sample is immersed. Then, the pressure in the chamber is progressively increased while the volume of intruded mercury in the pores of soil is recorded. The diameter of the intruded pores at certain pressure is obtained from the Young-Laplace equation (Eq. (3.2), Chapter 3), using the appropriate parameters of surface tension for the air-mercury interface and the contact angle between mercury and solid particles. Finally, a graph of the relative intruded volume *versus* the size of pores is produced. With these results, the sizes of macropores and mesopores can be established. However, the unrealistic hypotheses made to determine the pore sizes, together with the impossibility of measuring the whole range of sizes [46], as well as doubts related to the deleterious effect of high mercury pressures on the size of pores for loose soils [65] in addition to some inconsistencies on the application of data to correctly

Eduardo Rojas

reproduce the soil water retention curves (SWRCs) [66], require these results to be taken with caution and to be considered only as an approximation to the real PSD of the material.

5.2. MERCURY INTRUSION POROSIMETRY TESTS

Recently, the use of the MIP test to ascertain the PSD of soils has become quite popular in unsaturated soil mechanics, primarily because of its simplicity. To perform this test, a sample of around 1 cm^3 is introduced into a cell filled with mercury. The sample has been previously dried by means of different techniques, being two of them the most frequently used: oven-drying and freeze-drying. In general, the freeze-drying method is preferred as it is associated with a smaller disturbance of the original structure of the soil due to the rapid rate of freezing. Once the sample has been placed in the cell, the pressure of mercury is gradually increased and the volume of intruding mercury is measured. The radii of the intruded pores are obtained from the Young-Laplace equation involving the superficial tension of mercury and the contact angle between mercury and soil. The main hypothesis employed to interpret these results is to assume that only those pores having the size of a critical radius (determined from the Young-Laplace equation for the current mercury pressure) are intruded at each increment of mercury pressure. Then, a graph showing the relative volume of pores (in cm^3 per gram) for each pore size is produced. More details of the equipment and procedure required to obtain the PSD of soils by MIP are reported by Simms and Yanful [45].

However, the hypothesis made to interpret MIP tests is clearly unrealistic. It supposes that only equally sized pores are interconnected, while there is no interconnection between pores of different sizes. In fact, it has been acknowledged that the results of MIP tests exaggerate the frequency of small pores, while underestimating that of large pores [67]. This is a result of the intrusion of mercury in the bonds, matching that of the cavity connected to this bond. This is explained by the fact that larger pores are the first ones to be intruded when mercury pressure increases. Therefore, when a bond is intruded, the cavity connected to this bond is also intruded.

Considering a porous-solid model as the one described in Chapter 3, it is possible to give a better interpretation of the results of MIP tests. The invasion of mercury (which is a non wetting fluid) is similar to a drying process where pores are invaded by air (which is also a non wetting fluid), forcing the water to drain out of the sample. In both cases, the largest pores are the first ones to fill in with fluid as indicated in Fig. (**1**).

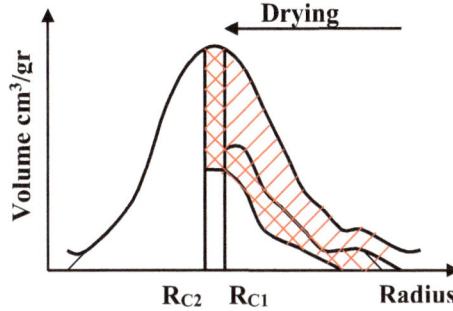

Fig. (1). Pores invaded by mercury (shaded zone) during a MIP test.

When mercury pressure increases, only a fraction of pores whose size is equal or larger than the critical radius will saturate, while the rest will remain blocked by smaller bonds. This occurs because of the interconnection of pores of all sizes. Some larger pores saturate during this increase of mercury pressure because at least one of their interconnected bonds belongs to those that saturate during this last increment.

The mechanism of mercury invasion is sketched in Fig. (**1**). Considering that the critical radius reduces from R_{C1} to R_{C2} due to the increase in mercury pressure, the blank zone in the figure represents the volume of pores that still have not yet been invaded by mercury. The single shadow zone represents the volume of pores already invaded by mercury before the new pressure increment. Finally, the double shadow zone represents the volume of pores filled in during the new pressure increment.

For MIP tests, the appropriate values of contact angle and surface tension are needed. According to Eq. **3.2**, the ratio between the suctions in a pore filled with water (S_w) and mercury (S_m) is given by the relationship

$$\frac{S_w}{S_m} = \frac{T_{sw} \cos \theta_w}{T_{sm} \cos \theta_m} \tag{5.1}$$

where T_{SW} and T_{Sm} represent the surface tension for water and mercury, respectively and θ_w and θ_m are the contact angles for water and mercury with the soil minerals, respectively.

Therefore, the porous-solid model can be used to simulate MIP tests by considering that the sample is subjected to a drying process, in which the air volume represents the intruded mercury and the critical radius is computed by means of Eq. (3.2).

As stated previously, the use of MIP tests on plastic soil has been questioned [65], and for that reason the evaluation of the capabilities of the porous-solid model to interpret MIP tests has been carried out with respect to the results reported by Roels, Elsen, Cermeliet and Hens [68] on a rigid calcareous sedimentary rock from Savonnières, France. These authors reported the porosimetry study of this material by using two different techniques: image analysis and mercury intrusion. For the image analysis, they obtained a series of micrographs of a cross-section of the rock by means of a scanning electron microscope (SEM). Then, by applying the spherical pore segment model, the authors could define the relative volume for each pore. Fig. (2) shows the results from both tests, where a marked difference between these two techniques can be observed. It is worth noting that MIP tests may report pore sizes up to two orders of magnitude smaller than the image analysis technique. Similar results have been reported when comparing the MIP method with other techniques [67].

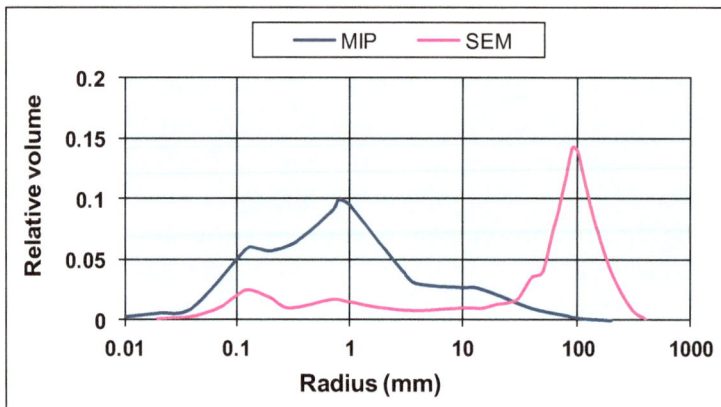

Fig. (2). Porosimetry results by MIP and SEM (after [68]).

Fig. (3) shows the wetting and drying SWRC of the limestone obtained from a pressure membrane apparatus. In this case, the material is stiff enough to avoid appreciable volume changes during wetting-drying cycles and therefore its void ratio remains approximately constant, meaning that, the phenomenon of progressive collapse of pores is not required in these simulations. Hence, it has been possible to relate the water content of the sample with its degree of saturation. On the other hand, it has been assumed that the PSD obtained from the image analysis technique reflects the real PSD of the rock and consequently was adopted as the distribution of cavities (macropores and mesopores) in the porous-solid model. Subsequently, a size distribution for bonds was proposed and the SWRC in wetting and drying was simulated and compared with the experimental results. The proposed distribution for bonds was then modified until the best fit for both curves was obtained. During the fitting process it was considered that the size distribution of bonds affects mainly the drying curve, while it shows a minor effect on the wetting curve, as aforementioned. In contrast, the size distribution of sites affects mainly the wetting curve, while it shows a minor effect on the drying curve. On a later section of this chapter, the procedure to fit numerical with experimental SWRCs is explained.

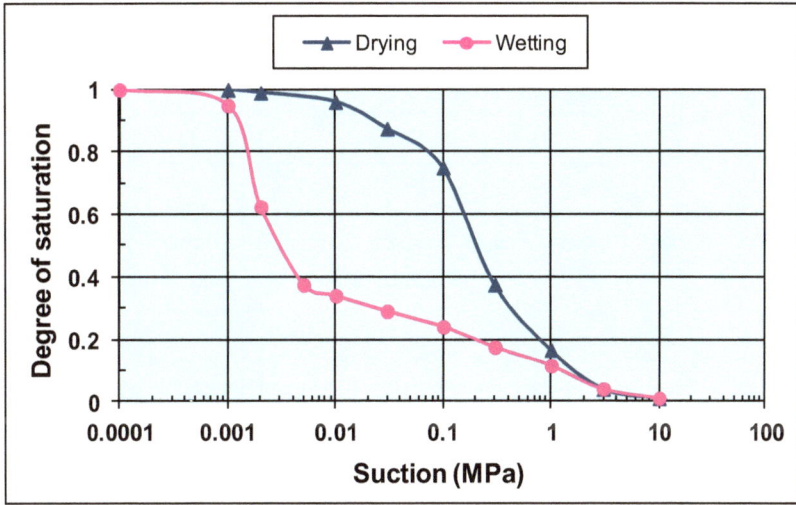

Fig. (3). Wetting and drying SWRC (after [68]).

The best fit for both curves is shown in Fig. (**4**). To reach this result a double logarithmic normal size distribution was adopted for bonds. The pore size density function for sites and bonds is shown in Fig. (**5**). The data for the adopted distribution for bonds is: $R_{B1} = 0.007\mu m$, $R_{B2} = 0.05\mu m$, $R_{B3} = 1.0\mu m$, $\delta_{B1} = 4$, $\delta_{B2} = 5.3$, $\delta_{B3} = 3$, $P_{vf1} = .025$, $P_{vf2} = .0015$.

Once the size distribution for bonds has been established, it is possible to simulate MIP tests using the considerations previously mentioned. The parameters required to simulate this test were reported by Roels, Elsen, Cermeliet, and Hens [68] which include: $T_{SW} = 0.073$ N/m, $T_{Sm} = 0.485$ N/m $\theta_w = 0$ and $\theta_m = 140°$. According to these parameters and Eq. (**1**), the equivalent water suction for this test is $S_w = 0.2s_m$. With this value, Fig. (**6**) was obtained. This figure shows the comparison between the experimental results of the MIP test with those obtained from the porous-solid model. This figure also shows the numerical PSD resulting from the density functions shown in Fig. (**5**). Even if some differences between the numerical and experimental MIP curves subsist, especially those related to the relative volume of small pores, the shape of both curves is very similar. It can also be verified that the size distribution of pores obtained from the simulation of the MIP test, reduces by two orders of magnitude the original PSD of the soil as it is observed in the experimental results shown in Fig. (**2**).

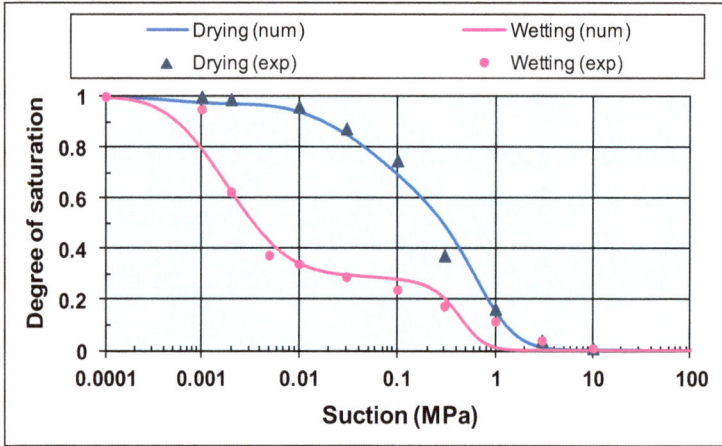

Fig. (4). Numerical and experimental wetting and drying SWRC (experimental data from [68]).

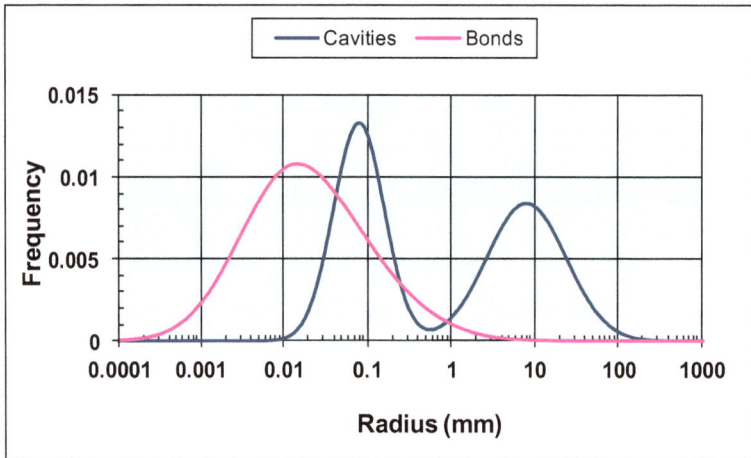

Fig. (5). Frequency function adopted for bonds and comparison with sites.

This result has serious implications regarding some procedures recently developed to define the SWRC or the hydraulic conductivity of unsaturated soils based on the PSD obtained from MIP tests. Finally, another important remark is that the PSD obtained from a single MIP test provides no sufficient information to define the size distribution of the two different porous elements, namely the cavities and the bonds, of the model adopted herein. To do so, additional information is needed. For example, the results of intrusion tests require being complemented with retraction tests as is currently performed in the nitrogen adsorption-desorption method.

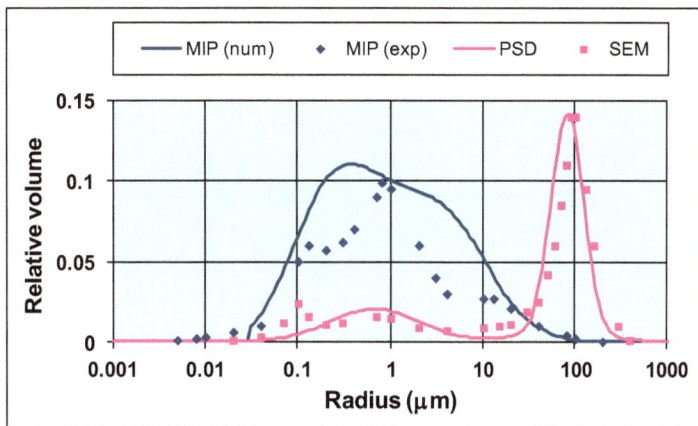

Fig. (6). Numerical and experimental comparison for a MIP test on a Savonnières rock (experimental data from [68]).

5.3. SOIL-WATER RETENTION CURVES

The SWRC is the relationship between suction and water content or degree of saturation of soils as explained before. The relevance of the SWRC in the study of unsaturated soils behavior has increased since the hydro-mechanical coupling phenomenon in these materials has been identified. Evidence of the hydro-mechanical coupling appears in the influence of the degree of saturation on the strength of unsaturated soils when identical samples are subjected to the same test conditions including suction. This means that a sample subject to wetting shows different strength than one following a drying path, even if suction is the same for both samples. Another evidence of this coupling effect is the influence of volumetric deformations (induced by mechanical actions on the soil sample) on the SWRC. Recent constitutive models for unsaturated soils involve the modeling of the SWRC to induce this hydro-mechanical coupling ([22, 24, 25]). Besides, recent developments in the design of pavements require using the SWRC [69]. Even the tensional strength of soils can be related to the SWRC [70].

Different methods and empirical relationships have been proposed to model the SWRC. Some of these equations use parameters related to the air entry value, the residual water content and the main slope of the curve ([71, 61]). Other methods are based on the GSD in addition to other soil properties, and use a statistical correlation between soil data and water content [72 - 74]. The degree of confidence of these models depends largely on the quality and quantity of the data used for the statistical correlation. Other methods use the PSD which, in some cases, is estimated from the GSD [75 - 77]. Otherwise, it is directly obtained from porosimetry tests [78, 79]. As the SWRC depends on the PSD of the material then, the precision in the PSD measurements affects the results derived from these models.

Likewise, different network models have been proposed to simulate the SWRC. For example, Androutsopoulos and Mann [80] proposed a 2D square network model made of randomly distributed cylinders of different sizes. The Young-Laplace equation was used to determine which pores can drain or saturate according to their size and current suction. Saturation or drainage starts at the borders and continues through the porous network in a quasi-static flow where suction monotonically reduces or increases. Considering a logarithmic normal distribution for the diameter of the cylindrical pores, these researchers were able to approximately reproduce the intrusion and retraction of mercury in a cobalt/molybdenum porous sample. However, the experimental PSD for this material was not reported and the differences with the numerical PSD used for the simulations rested unknown.

Simms and Yanful [79] used a similar network model that included the pore shrinkage phenomenon due to suction increase. With this model they tried to reproduce the experimental SWRC of different soils using the PSD obtained from MIP tests. Nevertheless, the comparison between numerical and experimental SWRCs for different materials was not very successful [78]. Similar results were reported by Zhang and Li [81]. The main reason for these poor results lies on the main hypothesis made to interpret MIP tests as discussed before.

Zhang and Li [81] performed a series of tests on five different mixtures of completely decomposed granite with sizes varying from gravel to clay. These tests included the GSD, the PSD and the drying SWRC. The PSD was obtained from MIP tests. The SWRC was achieved using a pressure plate apparatus for suctions up to 0.5 MPa. In some cases this method was complemented with the psychrometer technique reaching suctions up to 50 MPa. The basic parameters for the five mixtures of soil are shown in Table **1**.

The SWRCs for all five soils are presented in Fig. (**7**). Notice that the more the proportion of coarse material of the soil increases, the more the SWRC displaces to lower degrees of saturation. The only material that does not follow this tendency is the poorly graded sand (SP) which shows degrees of saturation similar to silty sand (SM).

Table 1. Basic parameters for the five different soils.

Soil Type	γ_{max} (g/cm³)	w_{opt} (%)	Clay (%)	Silt (%)	Sand (%)	Gravel (%)	ASTM D2487
1	1.97	9	3.9	3.1	16.7	76.3	GP
2	1.96	11	5.9	18.7	17.9	57.5	GW
3	1.90	13	7.9	34.3	19.1	38.7	SP
4	1.71	18	9.9	49.9	20.3	19.9	SW
5	1.55	21	11.9	65.5	21.5	1.1	SM

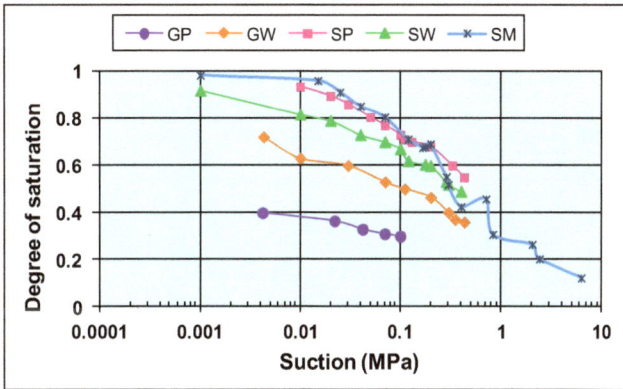

Fig. (7). Experimental SWRCs for the five soils tested (adapted from [81]).

The numerical PSD was obtained by fitting the numerical with the experimental SWRC of each material. The procedure used to fit the numerical with the experimental results is presented as an example in the next section of this chapter. The PSD resulting from this procedure is considered to be the porous-solid model PSD. Similarly, the GSD was obtained from the simulation of the cumulative finer by weight gradation of each soil employing a single, double or triple logarithmic normal distribution depending on the material.

Once the SWRC was correctly reproduced, the resulting PSD of the soil was used to simulate a MIP test according to the principles outlined in the previous section of this chapter. Finally, the results of the simulated MIP tests were compared with the experimental results.

Figs. (**8a**) to (**12a**) show the simulation of the SWRC for GP, GW, SP, SW and SM soils from which the numerical PSD of each material was established. The parameters adopted for the PSD and the GSD for the different soils are presented in Table **2**.

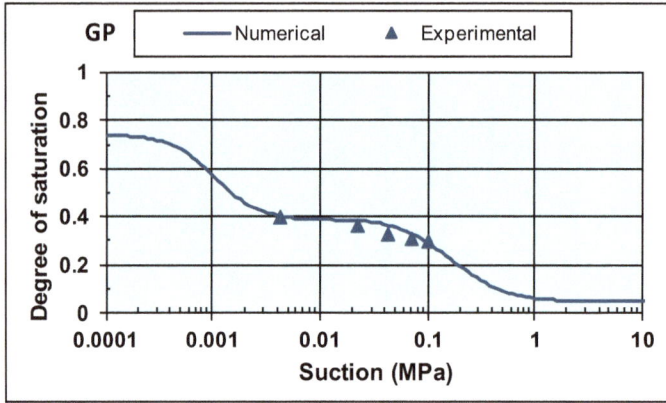

(a)

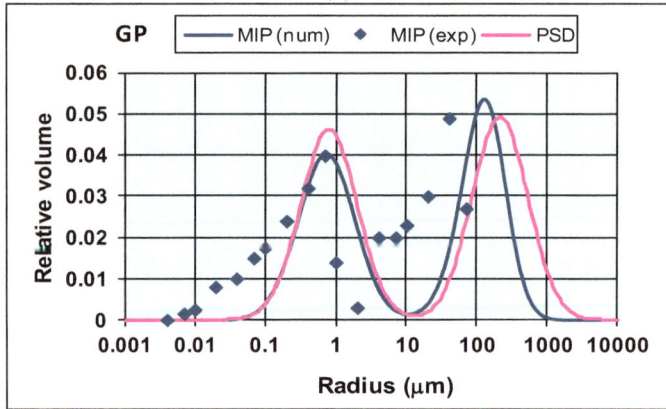

(b)

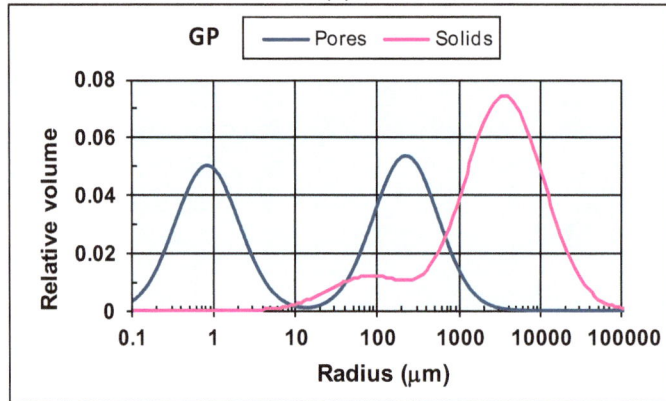

(c)

Fig. (8). Numerical and experimental results for (**a**) the drying curve, (**b**) the PSD and (**c**) numerical PSD with experimental GSD for GP soil (experimental data from [81]).

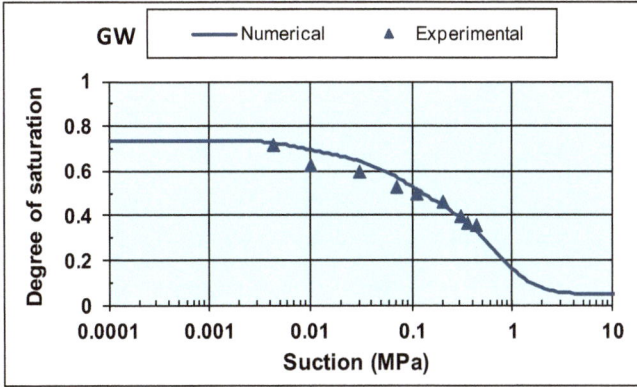

(a)

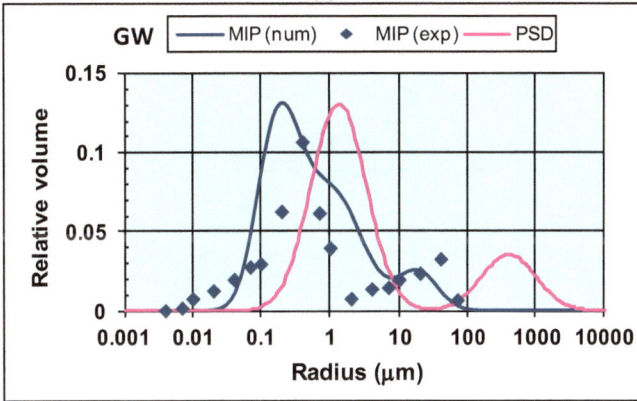

(b)

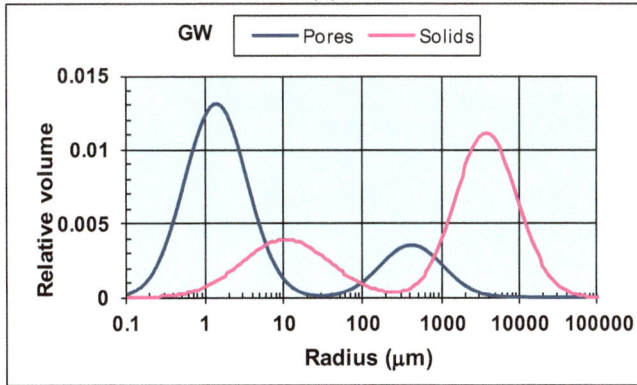

(c)

Fig. (9). Numerical and experimental results for (**a**) the drying curve, (**b**) PSD and (**c**) numerical PSD with experimental GSD for GW soil (experimental data from [81]).

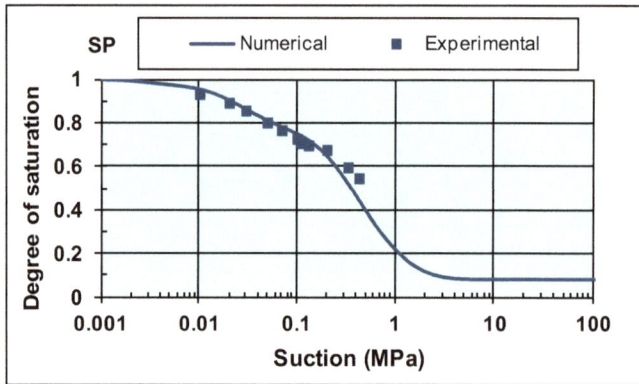

(a)

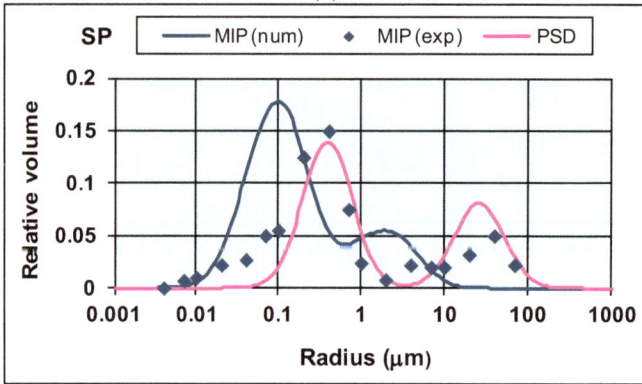

(b)

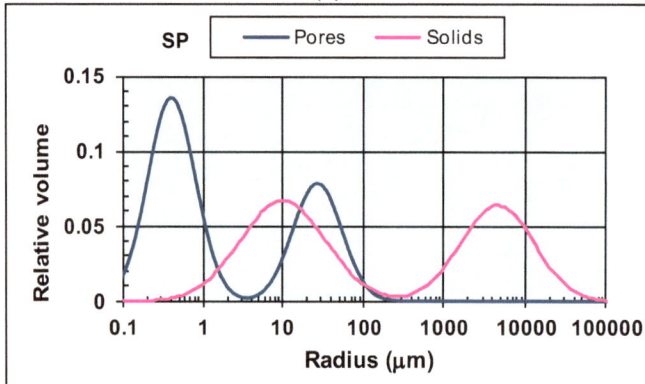

(c)

Fig. (10). Numerical and experimental results for (**a**) the drying curve, (**b**) PSD and (**c**) numerical PSD with experimental GSD for SP soil (experimental data from [81]).

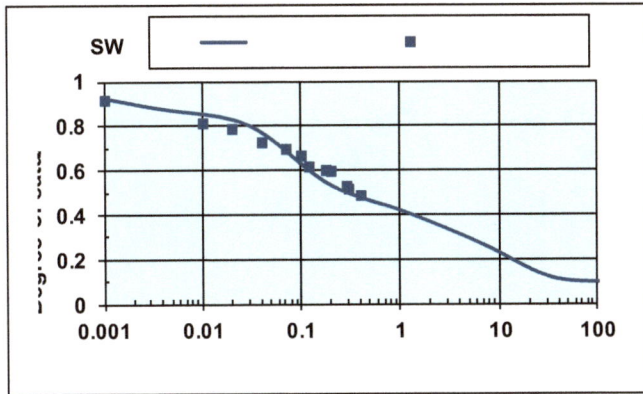

(a)

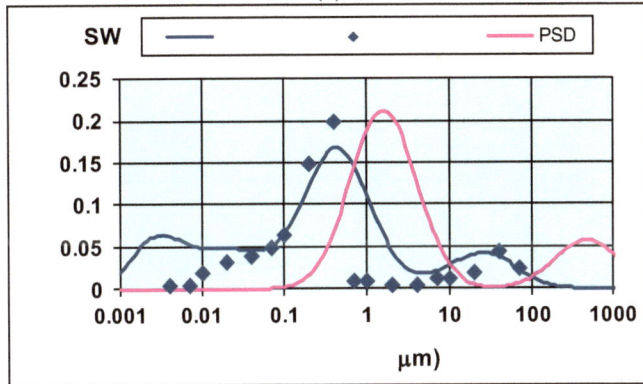

(b)

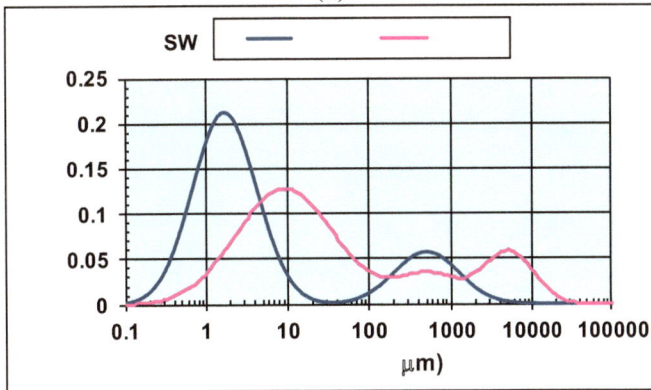

(c)

Fig. (11). Numerical and experimental results for (**a**) the drying curve, (**b**) PSD and (**c**) numerical PSD with experimental GSD for SW soil (experimental data from [81]).

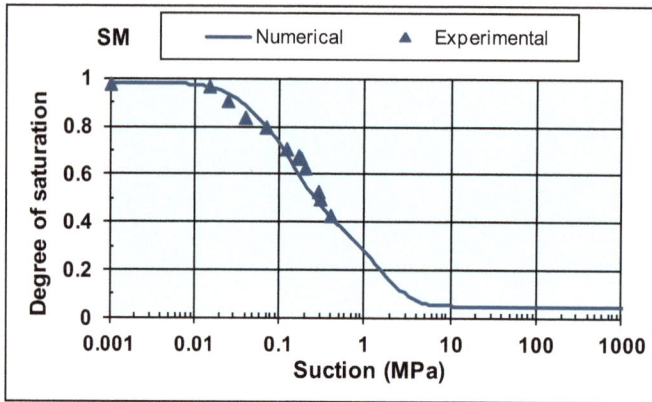

(a)

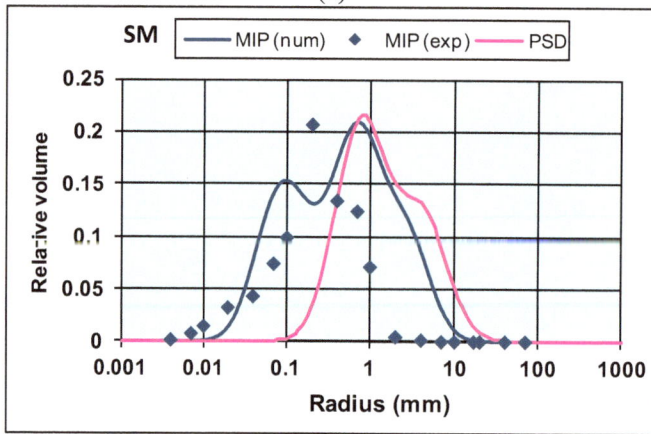

(b)

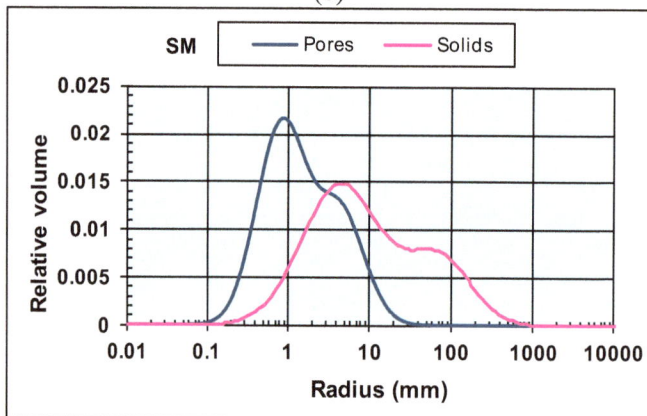

(c)

Fig. (12). Numerical and experimental results for (**a**) the drying curve, (**b**) PSD and (**c**) numerical PSD with experimental GSD for SM soil (experimental data from [81]).

Table 2. Mean size and standard deviation for the normal distributions for sites, bonds and solids of the materials described in Table 1.

Soil		S_1	S_2	B_1	B_2	B_3	Sol_1	Sol_2	Sol_3	e_0	S_f
GP	$\bar{R}$ *(μm)*	0.15	40	.04	25	0.0	7	320	0.0	0.705	.00015
	δ	2.5	2.5	5.0	3.0	0.0	3.0	3.0	0.0		
	P_{vf}		1.5E-5		.008			3E-3			
GW	$\bar{R}$ *(μm)*	0.15	15	.001	.005	0.0	0.4	700	0.0	0.61	0.061
	δ	2.5	2.5	1.5	3.0	0.0	3.6	2.5	0.0		
	P_{vf}		9E-6		.0005			3.3E-6			
SP	$\bar{R}$ *(μm)*	0.15	10.0	.015	1.0	0.0	0.5	550	800	0.62	0.017
	δ	2.0	2.0	3.0	3.0	0.0	3.4	2.8	2.0		
	P_{vf}		1.5E-4		6E-3			1.6E-6	1E-7		
SW	$\bar{R}$ *(μm)*	0.3	90	.001	.02	18	0.2	100	2000	0.63	0.0385
	δ	2.5	2.5	3.0	3.0	3.0	4.0	2.5	2.0		
	P_{vf}		3E-6		1E-2	1.5E-5		6E-6	4E-8		
SM	$\bar{R}$ *(μm)*	0.3	1.6	.01	1.0	0.0	0.4	13.0	0.0	1.02	0.144
	δ	2.0	2.0	3.0	3.0	0.0	3.0	2.5	0.0		
	P_{vf}		2E-2		3E-3			8E-4			

Notes: $\bar{R}$ = mean size, σ = standard deviation, S = sites (S), B = bonds, Sol = solids, sub-scripts from 1 to 3 indicate the three different logarithmic normal distributions, P_{vf} = Proportional volume factor, e_0 = initial voids ratio, S_f = shape factor

As can be observed in Figs. (**8b-12b**), the difference between the porous-solid model PSD and that obtained from the simulation of the MIP test is about one order of magnitude. This difference explains the failure of those network models that intend to reproduce the SWRC directly from the PSD obtained from MIP tests, and this arises from the fact that SWRCs are simulated with a network model while MIP tests are interpreted according to a model based on a bundle of capillary tubes.

The experimental PSD matches with the numerical results for all soils (see Figs. **8b**, **9b**, **11b** and **12b**) except for SP soil (Fig. **10b**) whose experimental points are closer to the porous network PSD. This result is caused by the fact that the experimental SWRC obtained for this material is very close to that of SM soil (see Fig. **7**), despite the fact that their experimental PSDs are quite different (compare Figs. **10b**, **12b**). In other words, the experimental results for SP soil cannot be correctly reproduced by the model because they show some inconsistencies.

It can also be observed in these figures that the numerical PSD of soils GP, GW,

SW and SM -obtained from the simulation of MIP tests- consistently show slightly smaller pore sizes than those experimentally determined, as in the case of the example presented in the previous section of this chapter.

The differences in the value of the contact angle when water or mercury intrudes or retracts from pores [55, 82] may explain this result, as the advancing angle has been found to be significantly larger than the receding one while the model considers the same value for both cases. In spite of this, it can be concluded that the probabilistic model is able to approximately define the PSD of the soil based on the SWRCs or *vice versa*.

It can also be noticed that all soils show a significant correlation between their numerical PSDs with their GSDs (see Figs. **8c-12c**), as it was experimentally observed by Alonso *et al.* (2008) for different soils. This result explains why some methods based on the GSD of the material have relative success in predicting the SWRC for soils (see for example [72, 75, 77]).

5.4. OBTAINING THE PORE SIZE DISTRIBUTION

The procedure to fit the numerical with the experimental SWRCs is shown in this section. At the end of this fitting process, the PSD of the soil is determined because, as stated before, there is a one to one relationship between the PSD and the SWRCs of porous materials.

The data published by Roels, Elsen, Cermeliet and Hens [68], used previously in the example presented in section 2 of this chapter, is also employed for this case. These researchers reported both the wetting and the drying SWRCs, the results of MIP tests and the PSD obtained from SEM analysis. For this instance, it is considered that the only available data are the main SWRCs of the soil. Therefore, numerical and experimental PSDs are used here merely for comparison during the fitting process.

The first step to adjust the numerical and the experimental SWRCs of a soil is to arbitrarily choose the initial values for the mean size and the standard deviation of pores. The larger pores (cavities) regulate the wetting curves while the smaller (bonds) regulate the drying curve, as explained before. When the SWRC shows a development, where suction varies in several orders of magnitude or when the PSD shows two or three peaks, then a double or triple logarithmic normal distribution is required. Suppose that the drying SWRC has already been fitted using a triple logarithmic normal distribution. This distribution was chosen because the SWRCs show a development in the suction axis of several orders of magnitude. As usually both the wetting and the drying branches require the same order of size distribution (single, double or triple), then a triple logarithmic normal distribution is also proposed to fit the wetting curve. The results obtained with the initially proposed values for this distribution (Table **3**) are shown in Fig. (**13**). Fig. (**13a**) shows the comparison between the numerical and experimental SWRCs while Fig. (**13b**) shows the proposed PSD compared with the SEM analysis and also the numerical and experimental PSD obtained from a MIP test.

Fig. (**13a**) shows the difference in slopes between the numerical and experimental wetting curves. When the numerical curves display steep slopes with regards to the experimental results, then the standard deviation needs to be increased. If the slopes are too smooth with respect to experimental results then the standard deviation needs to be reduced. In this case, the slopes of the numerical curve are too steep compared to experimental results, especially for large values of suction. This means that the standard deviation for the small cavities (Cavities 1 and 2) needs to be increased as indicated in Table **4**.

Table 3. Initial proposed values for cavity parameters considering a triple logarithmic normal distribution.

Element	Cavities 1	Cavities 2	Cavities 3
$\bar{R}$ *(μm)*	0.15	15	150
δ	1.5	1.5	1.5
P_{vf}		2E-5	2.5E-5

Table 4. Second try for cavity parameters.

Element	Cavities 1	Cavities 2	Cavities 3
$\bar{R}$ *(μm)*	0.15	15	150
δ	3.0	3.0	1.5
P_{vf}		2E-5	2.5E-5

Fig. (**14**) shows the results obtained with the values indicated in Table **4**. A better correlation of slopes for the numerical and experimental SWRCs can now be observed in Fig. (**14a**). However, the curves look displaced to the right hand side. This implies that the mean size of cavities needs to be reduced in order to increase the suction required to intrude the pores. Therefore, the sizes of cavities are reduced as indicated in Table **5**.

The results obtained with the values of Table **5** can be observed in Fig. (**15**). A final adjustment of parameters for the best fit of the wetting curve is shown in Table **6**. These parameters result in the curves shown in Figs. (**4** and **5**) of this chapter. This same process can be applied to fit the drying curve. At this point it can be observed that the numerical PSD, obtained with the fitting process of the

SWRCs, is well correlated to the experimental PSD obtained from the SEM analysis. Besides the numerical and experimental PSD for a MIP test shows good correspondence (Fig. **6**).

Table 5. Third try for cavity parameters.

Element	Cavities 1	Cavities 2	Cavities 3
$\bar{R}$ *(µm)*	0.06	6	60
δ	3.0	3.0	1.5
P_{vf}		2E-5	2.5E-5

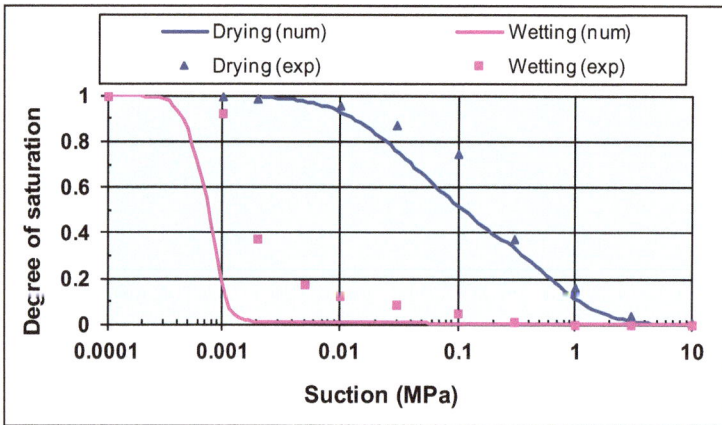

(a)

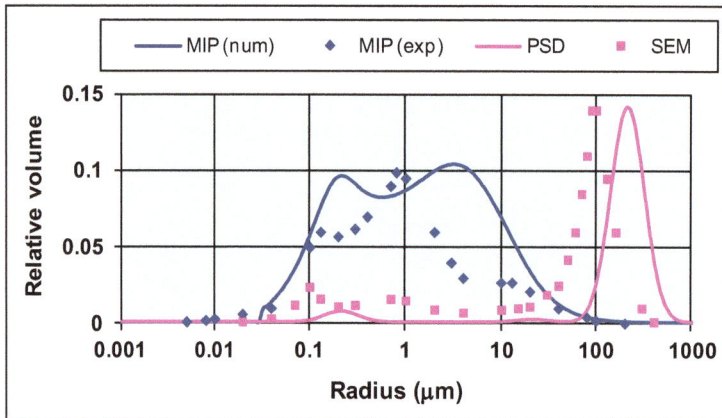

(b)

Fig. (13). Comparison between experimental and numerical results for (**a**) SWRCs and (**b**) PSD and MIP test for initially proposed parameters (experimental data from [68]).

Table 6. Final values for cavity parameters.

Element	Cavities 1	Cavities 2	Cavities 3
$\bar{R}$ *(μm)*	0.06	6	73
δ	3.0	3.0	1.4
P_{vf}		2E-5	2.5E-5

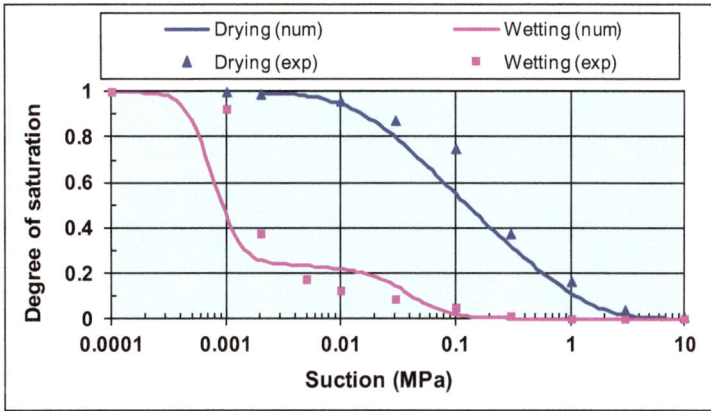

(a)

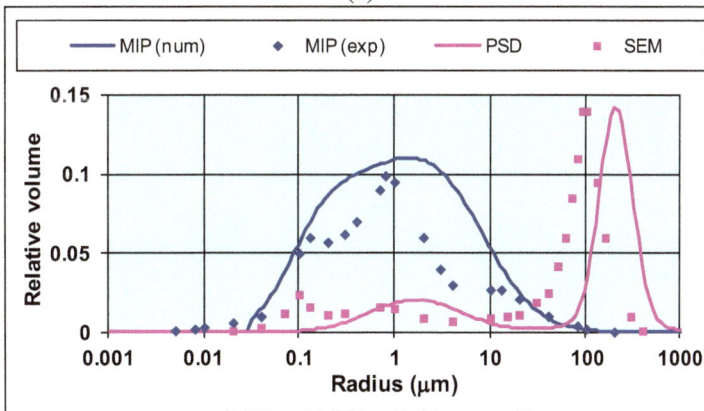

(b)

Fig. (14). Comparison between experimental and numerical results for (**a**) SWRCs and (**b**) PSD and MIP test. Second try (experimental data from [68]).

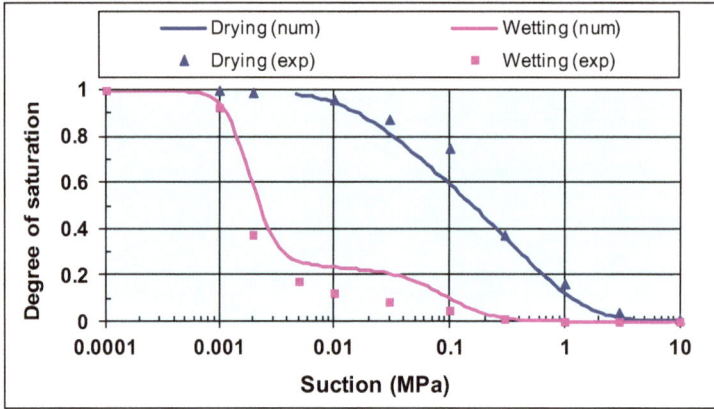

(a)

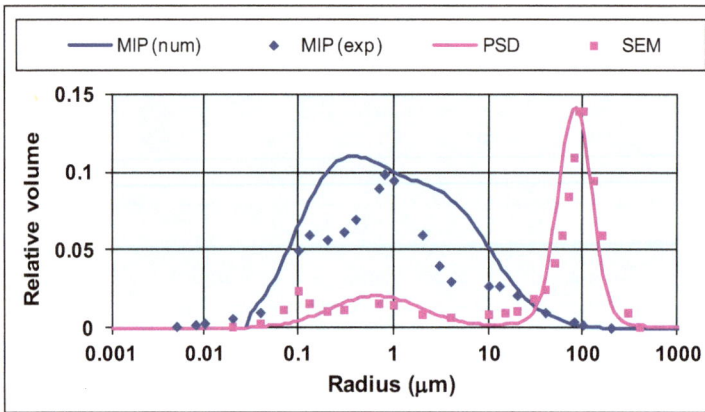

(b)

Fig. (15). Comparison between experimental and numerical results for (**a**) SWRCs and (**b**) PSD and MIP test. Third try (experimental data from [68]).

CHAPTER 6

Compression Strength of Soils

Abstract: In this chapter, the probabilistic porous-solid model is used to determine the mean effective stress of soils at failure. The plots of the deviator stress against the mean effective stress show a unique failure line for a series of triaxial tests performed at different confining net stress and suctions for both wetting and drying paths. This result confirms that the proposed effective stress equation is adequate to predict the shear strength of unsaturated soils. It also results in different strengths for wetting and drying paths as the experimental evidence indicates.

Keywords: Axis translation technique, Confining stress, Constant volume test, Critical state, Drying path, Eeffective stress, Friction angle, Grain size distribution, Logarithmic normal distribution, Net stress, Pore size distribution, Porous-solid model, Shear strength, Soil-water retention curve, Triaxial tests, Wetting path.

6.1. INTRODUCTION

The probabilistic porous-solid model can be used to obtain the mean effective stress at failure for a soil following any stress path. These results can be plotted against the deviator stress to determine the failure surface of the material. In this chapter, the experimental results of the Speswhite kaolin as reported by Wheeler and Sivakumar [83] are used. These researchers performed a series of triaxial tests with different stress paths. With these results, some points of the soil-water retention curve (SWRC) at wetting could be obtained. Also, the pore size distributions (PSDs) of samples statically compacted at different vertical pressures and water contents have been reported by Thom, Sivakumar, Sivakumar, Murray and Mackinnon [84]. Finally, the grain size distribution (GSD) of this material was reported by Espitia [85]. Because at this stage the probabilistic model does not consider volume changes, only those paths involving no volume change of the sample during shearing are considered for the numerical comparisons. For the same reason, the experimental results were considered in three different groups depending on the confining stress applied to the sample. These groups correspond to the confining pressures of 0.1 (three tests), 0.2 (two tests) and 0.3 MPa (one test). Each group corresponds to a different PSD resulting in three different sets of

SWRCs and three different groups of curves for parameters f^s, f^d and S_w^u. Accordingly, numerical and experimental comparisons were made independently for each group.

6.2. NUMERICAL AND EXPERIMENTAL COMPARISONS

All samples used for the determination of the PSD, the SWRC and tested in triaxial tests were prepared by static compaction at a water content of 25% (4% less than the optimal). These samples were compacted in nine layers at a constant displacement of 1.5 mm/min and a maximum vertical total stress of 0.4 MPa. This procedure provided samples with dry density of 1.2 g/cm³, specific volume of 2.21 and 54% degree of saturation. Prior to the loading stage all samples were subject to an isotropic net stress of 0.050 MPa with suctions ranging from 0 to 0.3 MPa in the triaxial cell. At these levels of suction, all samples increased their water content. In addition, those samples subject to suctions of 0 and 0.1 MPa experienced volumetric collapse. Once equilibrium was accomplished, the isotropic net stress was increased to reach a final value ranging between 0.1 and 0.3 MPa. Because all samples increased their water content during the equilibrium stage, it is considered that all these results correspond to the wetting branch of the SWRC.

The GSD of the Speswhite kaolin reported by Espitia [85] is shown in Fig. (1). The same figure shows the adjusted numerical curve obtained from a double logarithmic normal distribution. Even though, small differences between these two curves exist, the numerical fitting is sufficiently accurate.

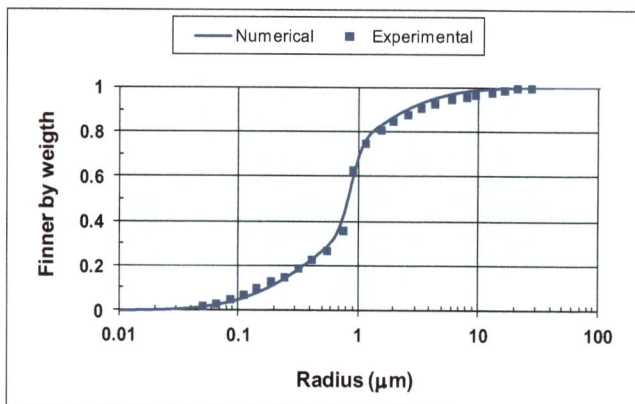

Fig. (1). Numerical and experimental GSD (experimental data from [85]).

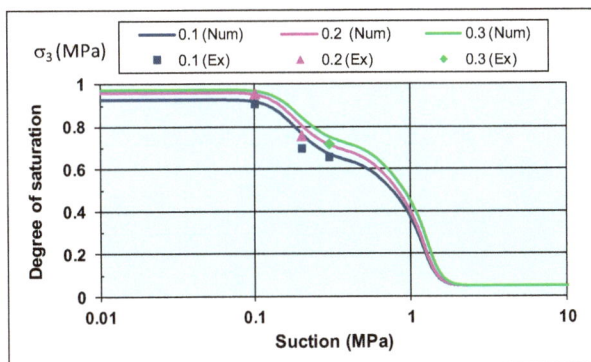

Fig. (2). Numerical and experimental SWRCs for different confining pressures (experimental data from [86]).

The experimental points of the SWRCs for the confining pressures of 0.1, 0.2 and 0.3 MPa are shown in Fig. (**2**). These points were obtained from the results of controlled suction triaxial tests at no volume change performed by Sivakumar [86]. They correspond to the value of the degree of saturation at critical state for those tests performed at the same confining pressure but different suction. This figure also shows the numerical SWRCs at wetting for the different confining pressures. The numerical curves were fitted to the experimental points by successively modifying an initially proposed PSD according to the procedure outlined in the previous chapter. In order to produce complete curves, it was necessary to estimate the value of the residual and the saturated degree of saturation according to the tendency of the experimental points. The first parameter was assessed as 0.05 for all tests, while the second was estimated in 0.91, 0.96 and 0.97 for the confining pressures of 0.1, 0.2 and 0.3 MPa, respectively.

Fig. (**3**) shows the PSD obtained from mercury intrusion porosimetry (MIP) tests carried out on a sample prepared according to the aforementioned procedure. This curve shows a bimodal distribution with two peaks: one at approximately 0.45 μm and the other at approximately 4.5 μm, corresponding to the size distribution of mesopores and macropores, respectively. The same figure shows the PSDs obtained by fitting the numerical and the experimental SWRC at wetting for the three different confining stresses. Although the experimental and the numerical curves show similar shapes, two main differences between them emerge: the first one is that the numerical PSD is displaced to the left with respect to the experimental results. The second one is that the numerical maximum relative volume of macropores is much smaller than the experimental value. The reason for these differences can be explained by the fact that MIP tests were performed in "as compacted" soil samples before the equalization stage where the confining pressure and the increase in water content produced a volumetric reduction in the sample which affects mainly the size of macropores as has already been discussed in chapters 3 and 5. This same deviation of the numerical PSD with respect to the experimental results was observed when a computational model was used to simulate the SWRC of this material [87].

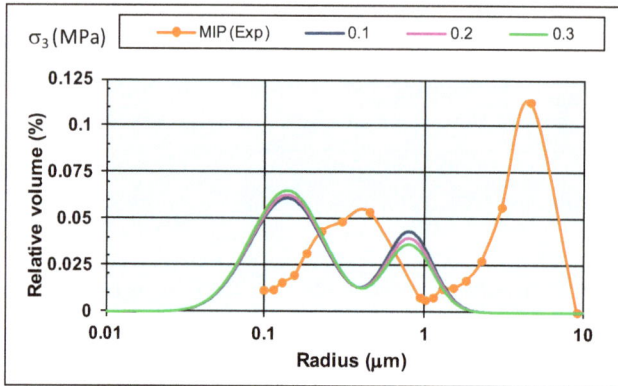

Fig. (3). MIP test results and numerical PSDs for different confining pressures (experimental data from [84]).

Once the PSD for each confining pressure has been established, the determination of the parameters of the porous-solid model is completed. Table **1** shows the parameters obtained for a confining pressure of 0.1 MPa. Notice that the bonds needed a double logarithmic distribution (B_1 and B_2) to better simulate the SWRCs.

Table 1. Parameters of the model for a confining pressure of 0.1 MPa.

Parameter	M	S	B_1	B_2	Sol_1	Sol_2
Mean size $\overline{R}$ (μm)	0.8	0.05	0.13	.008	1.1	.02
Standard deviation δ	1.4	1.7	1.4	1.7	1.1	3.8
P_{vf}	0.01		0.01		.005	

Notes: M = macropores, S = mesopores, B_1 = Bonds (1), B_2 = Bonds (2), Sol_1 = Solids (1), Sol_2 =Solids (2), P_{vf} = Proportional volume factor

For the confining pressures of 0.2 and 0.3 MPa all parameters included in Table **1** remain the same except for the mean size of macropores which take the value of 0.83 μm and 0.75 μm, respectively. The connectivity was considered to be 4. Finally, a value of 0.25 for the shape factor allowed matching the numerical and experimental voids ratio for the different confining pressures as shown in Table **2**.

Table 2. Experimental and numerical voids ratio for different confining pressures.

Confining stress (MPa)	0.1	0.2	0.3
Experimental voids ratio	1.18	1.13	0.99
Numerical voids ratio	1.17	1.15	1.0

Once all parameters of the porous-solid model have been defined, it is possible to simulate a wetting process and obtain the values of f^s, f^d, S_w^u and χ for the full range of suction and for each confining pressure. These results are presented in Fig. (**4**). In Figs. (**4a** and **4c**) it can be seen that both parameters f^s and S_w^u increase continuously with the degree of saturation although the first parameter grows at an increasing rate, while the second at a decreasing rate. On the contrary, Fig. (**4b**) shows that parameter f^d initially increases up to a maximum and then decreases with the degree of saturation. Fig. (**4d**) shows the numerical and experimental values of parameter χ for each confining pressure. The numerical values were derived from Eq. (2.15) (Chapter 2), whereas the experimental ones were obtained by assuming that the failure surface is represented by a single line in the axes of mean effective stress against deviator stress as expressed by Eq. (2.19) (Chapter 2). It can be observed that the numerical and experimental results lie fairly close for the confining pressures of 0.1 and 0.2 MPa. A similar comparison can be made for the matric suction stress σ_s^* as defined in Eq. (2.17), Chapter 2. This comparison is presented in Fig. (**5**) and again experimental and numerical results are quite similar for all confining pressures. The numerical results indicate that the maximum strength of the soil can be reached at a degree of saturation of 0.5 or 0.6 which seems rather low. In order to improve this prediction both branches of the SWRC for the entire range of suction would be required.

Finally, Fig. (**6**) shows the experimental results on the effective stress plane. The effective stress for the experimental points was obtained from Eq. 2.14 using the corresponding value of χ for each sample. This value can be obtained from the numerical curves shown in Figs. (**2** and **4d**) according to the suction and confining pressure applied to each sample. It can be observed that tests performed at different suctions and confining stresses align in a single failure surface.

As SWRCs are linked to the current porosimetry of the soil, any action affecting the last modifies the former. Therefore, a porous-solid model intending to represent the distribution of water into the soil should include the pore size change generated by loading or suction increase. According to Simms and Yanful [46, 78], macropores are responsible for most of the volumetric strain of soils. These results indicate that macropores keep reducing their size with further increments of load or suction up to the point where most of them reach the range size of

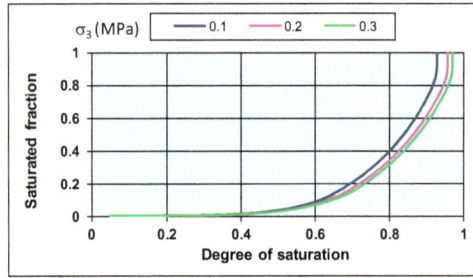

(a)

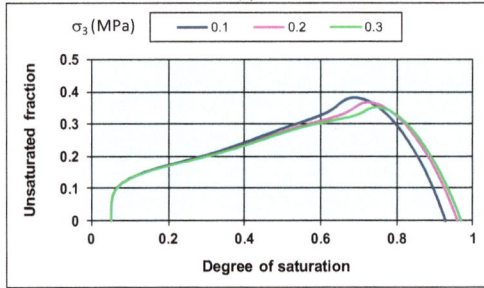

(b)

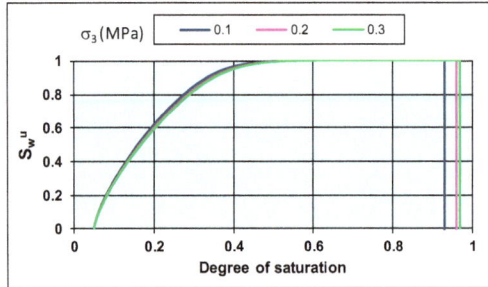

(c)

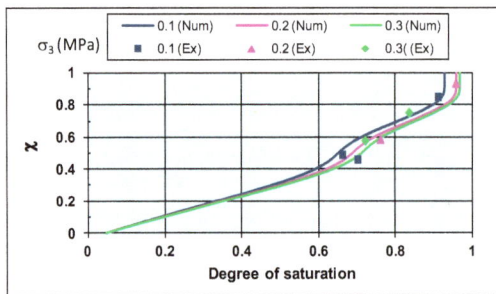

(d)

Fig. (4). Parameters derived from the porous-solid model for different confining pressures (**a**) saturated fraction $f^{\overline{s}}$, (**b**) unsaturated fraction f^u, (**c**) degree of saturation of the unsaturated fraction S_w^u and (**d**) parameter χ (experimental data from [83]).

mesopores. Still, most mesopores basically maintain their original size during loading or suction increase. This type of behavior can be introduced into the model by proportionally reducing the size of macropores with the level of loading as proposed by Koliji, Laloui, Cusinier and Vulliet [88]. In such cases the porosimetries of the sample before and after the shear test would be required.

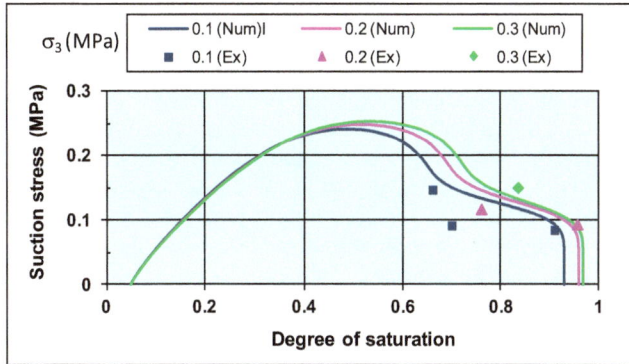

Fig. (5). Numerical and experimental matric suction stress for different confining pressures (experimental data from [83]).

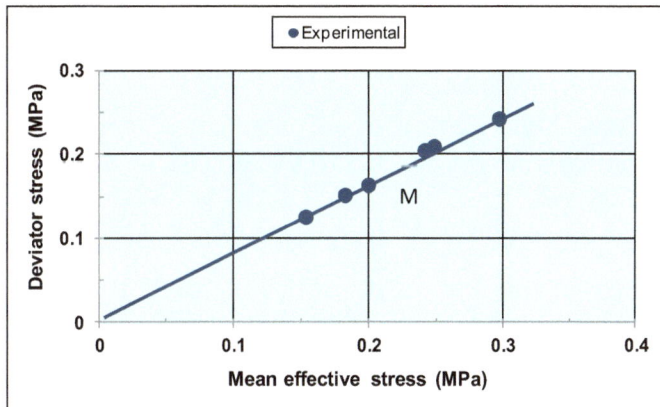

Fig. (6). Experimental results on the effective stress plane (experimental data from [83]).

CHAPTER 7

Tensile Strength

Abstract: In this chapter, the probabilistic porous-solid model is used to simulate the tensile strength of unsaturated soils tested at different water contents. The strength of unsaturated soils can be split in two parts: one related to the net stress and the other to suction. The strength generated by suction has its origin on the additional contact stresses induced to solid particles by water meniscus. This additional contact stress is called matric suction stress. In that sense, the tensile strength of soils represents the matric suction stress of the material at that particular water content. The numerical and experimental comparisons of the tensile strength of unsaturated soils tested at different water contents show that the probabilistic porous-solid model can simulate this phenomenon with sufficient accuracy.

Keywords: Additional contact stress, Cohesion, Direct tensile test, Effective stress, Homogenous material, Matric suction stress, Net stress, Probabilistic porous-solid model, Retention curves, Suction, Tensile strength, Tensile stress, Water menisci.

7.1. INTRODUCTION

Eq. (2.18) represents the strength of an unsaturated soil subjected to certain suction. This equation can also be rewritten as

$$\tau = \sigma'_n \tan \varphi = \left((\overline{\sigma})_n + \sigma_s^* \right) \tan \varphi = (\overline{\sigma})_n \tan \varphi + c$$

where c represents the cohesion of the soil. If osmotic suction is neglected, the matric suction stress represents an additional contact stress induced by water meniscus to solid particles (Lu, 2008) and is given by the relationship

$$\sigma_s^* = \chi s = \left[f^s + S_w^u f^u \right] s \tag{7.1}$$

Between the hypothesis made to obtain this equation is that the soil is considered as a homogeneous isotropic material and in that sense the matric suction stress represents an isotropic stress. During a pure tensile test the maximum strength reached by a soil sample represents the linking stress between solid particles and therefore, it also represents the matric suction stress of the material at that particular water content [89]. Thus, tensile tests can give a direct value of the matric suction stress of soils. In that sense, the probabilistic porous-solid model can be used to determine the matric suction stress of a soil and therefore its tensile strength.

7.2. TENSILE TESTS

Vesga and Vallejo [89] performed a series of direct tensile tests on kaolin samples with different degrees of saturation following a drying path. At the same time, these researchers reported the soil-water retention curve (SWRC) of the material.

The tensile tests were performed on flat bowtie-shaped samples. In this way, the samples could be fixed at their extremes and the failure always occurred at their center. The samples were 7 cm long, 2.2 cm thick with a central neck 2.5 cm wide. These samples were casted in a flat mould where the material was placed at a water content close to the liquid limit (40%). Then a vertical load of 0.03 MPa was applied for 24 hours. Once the loading stage was finished, the sample was subjected to a drying process in controlled humidity conditions up to the point where it reached a water content previously specified. Finally, the sample was placed in a membrane for 48 hours to allow the homogenization of the humidity before the test was performed.

Unfortunately, all these tests were performed following a drying path and there is no information related to the wetting path. Nevertheless, the porous-solid model was used to simulate the SWRC of the material by successively adjusting an initially proposed pore size distribution (PSD) as already explained in Chapter 5. Fig. (**1a**) shows the experimental SWRC obtained by Vesga and Vallejo [89] using the filter paper method. This figure also shows the fitted numerical SWRC obtained with the porous-solid model. In this case, a single logarithmic function was considered for both macropores and mesopores, whereas a double logarithmic distribution was considered for bonds to achieve the best fit for the SWRC. The required data for each distribution are the mean radius, the standard deviation and the proportional volume factor. These values obtained for these parameters are presented in Table **1**.

Table 1. Parameters of the model.

Parameter	M	S	B_1	B_2
Mean size $\overline{R}$ (μm)	.075	0.0014	0.0009	.03
Standard deviation δ	1.5	5.1	7	3.5
P_{vf}	0.02			0.1

Note: M = macropores, S = mesopores, B_1 = Bonds (1), B_2 = Bonds (2), Sol_1 = Solids (1), Sol_2 =Solids (2), P_{vf} = Proportional volume factor.

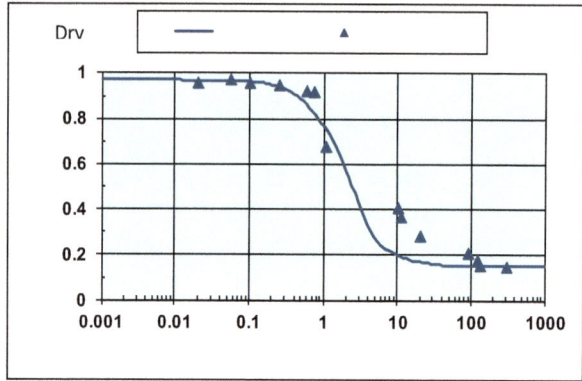

(a)

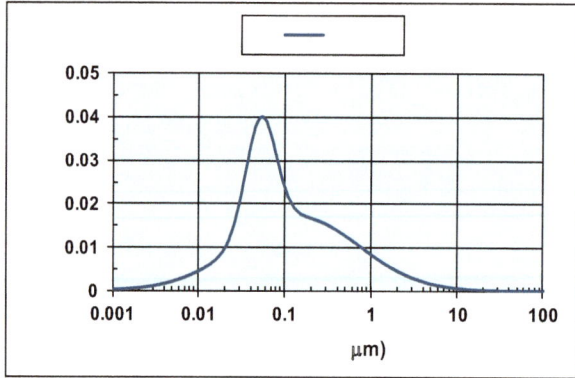

(b)

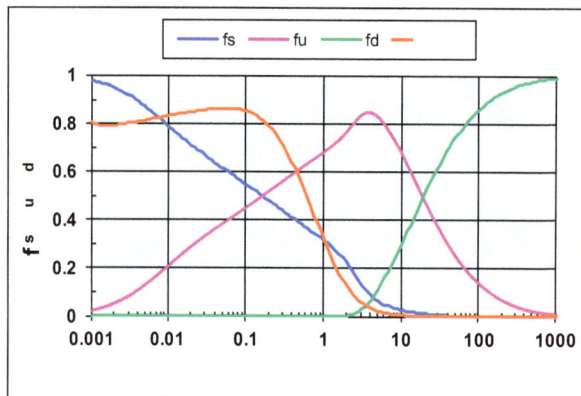

(c)

Fig. 1 cont.....

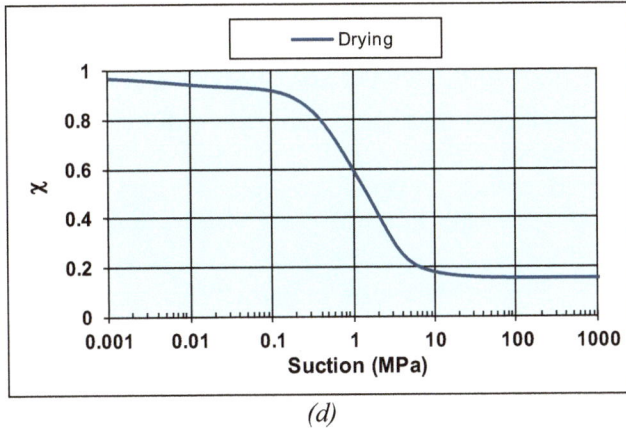

(d)

Fig. (1). Results for a kaolin sample: (**a**) Fit of the experimental SWRC, (**b**) Numerical PSD, (**c**) parameters f^s, f^u, f^d and S_w^u and (**d**) values of χ (experimental data from [89]).

These parameters establish the frequency of the different sizes of pores in the porous network. With this data and the size of pores, it is possible to determine the numerical relative volume for each size as shown in Fig. (**1b**). Fig. (**1c**) shows the values of parameters f^s, f^u, f^d and S_w^u obtained from the porous-solid model when the sample follows a drying path. Finally, Fig. (**1d**) shows the values for parameter χ *versus* the value of suction. By comparing Figs. (**1a** and **d**) it can be observed that the values of parameter χ are slightly smaller than those of the degree of saturation. All these parameters were obtained only for the drying condition as no information was provided for the wetting branch of the SWRC.

Fig. (**2**) shows the comparison between the experimental tensile strength and the matric suction stress obtained from Eq. (7.1). Because this last value represents the bonding stress between solid particles, it is equivalent to the tensile strength of the soil.

The numerical and experimental results presented in Fig. (**2**) show that a maximum tensile stress occurs at certain point of the drying process. This maximum is related to a maximum in the matric suction stress represented by Eq. 7.1. This maximum can be explained as follows: when an initially saturated soil is subjected to drying, the number of menisci producing the link between solid particles increases at the same rate as the unsaturated fraction increases (see Fig. **1c**). Then at certain point, the unsaturated fraction starts decreasing with increasing suction, meaning that the number of menisci decrease at certain point of the drying process. This occurs because at this point, a dry fraction develops and progresses with the value of suction as it can be observed in Fig. (**1c**). This reduction in the number of water menisci in the soil sample eventually results in a reduction of the matric suction stress even if suction keeps increasing.

The comparison presented in Fig. (**2**) shows that the model predicts a maximum tensile stress slightly larger than the experimental value. Additionally, the numerical maximum stress is displaced to the left hand side with respect to the maximum experimental value, however, the shape and values of the numerical curve correspond well with the experimental results. One problem with these comparisons is that the tensile tests were not performed in controlled suction conditions, while the numerical results consider that suction remains constant during the test.

In any case and according to these results, it can be said that the probabilistic porous-solid model simulates with fair precision the results of tensile tests performed at different water contents.

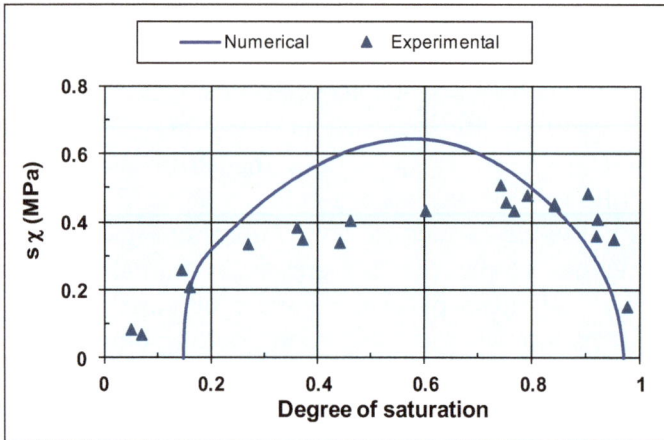

Fig. (2). Numerical and experimental results comparison for tensile tests (experimental data from [89]).

CHAPTER 8

Volumetric Behavior

Abstract: An equation to account for the volumetric behavior of unsaturated soils is proposed in this chapter. This equation is based on the effective stress principle and results in a unifying framework for the volumetric behavior for both saturated and unsaturated soils. The results of the proposed equation are compared with experimental results published by different researchers. These comparisons show that the equation is adequate to account for wetting-drying and net stress loading-unloading paths. This analysis confirms that the effective stress principle can be applied to the volumetric behavior of unsaturated soils.

Keywords: Collapse, Compression index, Controlled suction test, Effective stress principle, Effective stress, Elastoplastic framework, Hydro-mechanical coupling, Isotropic triaxial test, Macropores shrinkage, Suction hardening, Unloading-reloading index, Unsaturated soils, Volumetric behavior, Water menisci, Yield surface.

8.1. INTRODUCTION

Different approaches have been proposed to simulate the volumetric behavior of unsaturated soils. Two of the main trends are the independent stress variables approach and the single stress variable approach. In the first one two different coefficients are used to account for the contribution of net stress and suction on the volumetric behavior. In the second case, a single volumetric coefficient is related to a single stress variable (in most cases referred as the effective stress) to simulate the volumetric behavior.

One of the main advantages in using the single stress approach is that the hydro-mechanical coupling observed in unsaturated soils is implicit into the formulation. On the other hand, the difficulties in finding a correct explanation for the phenomenon of collapse upon wetting were one of the main objections to this approach. However, it is presently acknowledged that the simulation of this phenomenon requires, in addition to the effective stress equation, an appropriate elastoplastic framework. In contrast, the independent stress variables models seem to clearly explain the phenomenon of collapse upon wetting, while the implementation of the hydro-mechanical coupling has been included in different degrees ([22, 23, 90, 91]).

The first approach has the following general form for the elastoplastic volumetric strain increment $d\varepsilon_v$:

$$d\varepsilon_v = \frac{1}{v}\left(\lambda_{vp}\frac{d\bar{p}}{\bar{p}} + \lambda_{vs}\frac{ds}{(s+p_{atm})}\right)$$

Where v is the specific volume of the soil, $\bar{p}$ and $d\bar{p}$ represent the apparent preconsolidation mean net stress at the current suction and its increment, respectively, s and ds are the maximum previous suction and its increment, p_{atm} is the atmospheric pressure, λ_{vp} and λ_{vs} are the slopes of the compression curves due to increases of the mean net stress and suction, respectively, in a semilogarithmic plane. Both slopes show negative values meaning that a negative volumetric strain indicates volumetric reduction. This expression allows great flexibility in the simulation of the volumetric behavior of unsaturated soils. It is common to express λ_{vp} as a function of suction, while λ_{vs} is considered constant. However, the experimental results indicate that λ_{vp} must also depend on the mean net stress, while λ_{vs} must depend on both the mean net stress and suction (see for example [92] and [93]). In that sense, the above expression becomes more complicated than it seems. Another disadvantage of this expression is that under zero suction the equation for the volumetric behavior of saturated soils is not recovered and therefore there is not a smooth transition between saturated and unsaturated states [94]. Examples of this approach are given in the models developed by Alonso, Gens and Josa [7], Wheeler and Sivakumar [83] and Thu, Rahardjo and Leong [95], among others.

The second approach can be written in the following general form

$$d\varepsilon_v = \frac{\lambda_v}{v}\frac{dp'}{p'}$$

where p' and dp' represent the preconsolidation effective stress and its increment, respectively and λ_v represents the slope of the compression curve in the axes of the logarithm of the effective stress *versus* specific volume. If parameter λ_v is expressed as a function of suction alone, it shows decreasing values with increasing suction. This, however, contradicts the experimental results [94]. To avoid this inconvenient, λ_v should be written as a function of the mean net stress, the preconsolidation stress and suction. Another possibility is to write λ_v as a function of the degree of saturation [94]. The effective stress approach has been used in the models proposed by Sheng, Sloan and Gens [25], Sun, Cui, Matsuoka and Sheng [91], Khogo, Nikano and Miyazaky [96], Loret and Khalili [97], Kholer and Hofstetter [98] and Koliji, Laloui and Vulliet [99] among others.

Recently Sheng, Fredlund and Gens [100] proposed a combination of these two trends using two different volumetric parameters in conjunction with a stress parameter that accounts for the effects of both net stress and suction in the form

$$d\varepsilon_v = \lambda_{vp} \frac{d\bar{p}}{\bar{p}+s} + \lambda_{vs} \frac{ds}{\bar{p}+s}$$ (8.1)

Parameter λ_{vs} depends on the value of λ_{vp} according to the following relationship:

$$\lambda_{vs} = \begin{cases} \lambda_{vp} & s < s_a \\ \lambda_{vp} \frac{s_a+1}{s+1} & s > s_a \end{cases}$$ (8.2)

where s_a represents the saturation suction [100]. In this case, the volumetric strain by net stress or suction increase depends on both the current net stress and the current suction; therefore, Eq. (8.1) is able to reproduce more accurately the volumetric response of unsaturated soils reported in the international literature.

One of the most important features of this equation is the introduction to some extent of the hydro-mechanical coupling through parameter s_a. In addition, although the two compression indexes λ_{vs} and λ_{vp} can be related using Eq. (8.2), different approaches can be used for more general cases. When plotted in the mean net stress axis *versus* suction, the yield surface generated with Eq. (8.1) shows a concavity. In fact, most constitutive models for unsaturated soils show a concavity at the transition between saturated and unsaturated states (see for example [101, 102, 103 and 104]). Although, this concavity poses some difficulties in obtaining a unique response, this can be numerically solved. Moreover, Eq. (8.1) cannot be integrated and therefore requires special treatment in the stress integration of the constitutive model.

8.2. PROPOSED EQUATION

Juárez-Badillo [105] and Butterfield [106] proposed the following equation for the volumetric behavior of saturated soils

$$d\epsilon_v = \frac{dv}{v} = \lambda_v \frac{dp'}{p'}$$

Integration of the above equation results in:

$$\frac{v}{v_0} = \left(\frac{p_0'}{p'}\right)^{\lambda_v}$$

where p_0' represents the initial effective stress corresponding to a volume v_0 in the virgin consolidation line. If the mean effective stress p' becomes very large, the specific volume $v=1+e$ tends to zero which is clearly inconvenient. A more likely relationship would involve the void ratio instead of the specific volume, in the form of:

$$\frac{de}{e} = \lambda_e \frac{dp'}{p'}$$ (8.3)

where λ_e represents the slope of the compression line in a logarithmic plane of effective stress *versus* void ratio and because the voids ratio reduces with increasing effective stress it exhibits negative values. A similar expression was proposed by Sheng, Yao and Carter [107] for the volumetric behavior of sands upon isotropic loading. Integration of the above equation results in:

$$\frac{e}{e_0} = \left(\frac{p'}{p_0'}\right)^{\lambda_e}$$ (8.4)

Fig. (1) shows the plot of this equation in the axes of the logarithm of the mean effective stress *versus* void ratio for different values of the compression index λ_e and for an initial void ratio of 1.14 at a mean effective stress of 0.02 MPa. Most soils show values of the parameter λ_e ranging between -0.05 and -0.3 in which case, the volumetric behavior for stresses in the range of civil engineering interest (0.1 and 10 MPa) can be approached to straight lines, as it is commonly done.

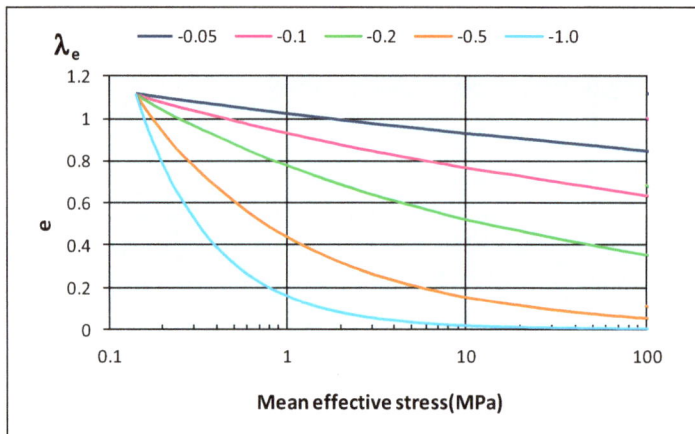

Fig. (1). Volumetric behavior of saturated soils for different values of the compression index.

For the case of unsaturated soils, its volumetric behavior can be analyzed through the results of two types of tests: isotropic loading by net stress increase at constant suction and suction increase at constant net stress. The results of the first type of test can be summarized as follows: when suction is below the air entry value, the soil behaves as saturated. However, when suction surpasses the air entry value, water menisci appear among the solid particles. These menisci produce additional contact stresses between the solids increasing the stability of the large pores as if the preconsolidation stress of the soil was increased. Therefore, the shrinkage of these pores can only be produced by applying further increments of the mean net stress. This means that the material experiences suction hardening. The greater the applied suction, the larger the mean net stress required for shrinking these pores and the soil behaves as an overconsolidated material.

For the second type of test, the soil behaves exactly the same as for the first case as long as suction remains below the air entry value. When suction becomes larger than the air entry value, a number of pores dry. Under these conditions, when suction increases, all saturated pores tend to shrink following the same law as for the saturated material. On the contrary, all dry pores will show no reaction to suction changes.

According to these descriptions, the evaluation of the volumetric response of unsaturated materials requires the quantification of the menisci of water and the pores that remain saturated at certain suction. For that purpose, the porous-solid model developed in Chapter 4 can be used.

The studies on the pore size distribution (PSD) of different soils indicate that in most cases these materials show a bimodal distribution meaning that they show two crests [78]: one corresponding to the macropores or large cavities and the other to the mesopores or small cavities. The macropores are cavities that show especial arrangements of solid particles in the form of vaults or arches. These pores have the characteristic of being larger than the solid particles forming the pore; therefore, their equilibrium is precarious upon isotropic loading or shearing. This can be confirmed by analyzing the PSD of different soils before and after performing triaxial tests as Futai and Almeida [92] and Simms and Yanful [45] have done. These researchers used the mercury intrusion porosimetry (MIP) test to compare the PSD before and after testing a soil. They found that most of the macropores reduce their size and transform into smaller pores at the end of these tests. They concluded that macropores are responsible for most of the volumetric response of soils.

On the other hand, the mesopores have the characteristic of being smaller than their surrounding solids; therefore, they are very stable and in general maintain their size upon isotropic loading or shearing. Sometimes soils show a mono-modal size distribution as for example uniform dense sands. In this case, macropores are absent from the soil and, in that case all volumetric deformation is generated by the shrinkage of mesopores which in any case is small compared with a bimodal structured soil. Mono-modal size distributions can also be obtained from slurries made of soils showing uniform grain size distribution (GSD). When slurries start to dry they adopt well defined structures with diverse PSDs which depend greatly on the GSD of the material. Bimodal GSDs usually derive on bimodal PSDs as it is observed in the experimental results shown in section 4 of this chapter. These bi-modal or mono-modal structures are also reflected on the shape of the soil-water retention curves (SWRCs). Samples with monomodal PSDs show a single inflection point in their SWRCs. On the contrary, samples with bimodal or trimodal PSD show two or three inflection points in their SWRCs, respectively. In any case, the porous-solid model can simulate any of these cases. For the monomodal case, it is sufficient to consider that macropores do not exist in the porous structure of the material.

The bonds represent the smallest pores usually found at the contact between solid particles and therefore they are also very stable at mean net stress or suction increments. All these elements are shown schematically in Fig. (2).

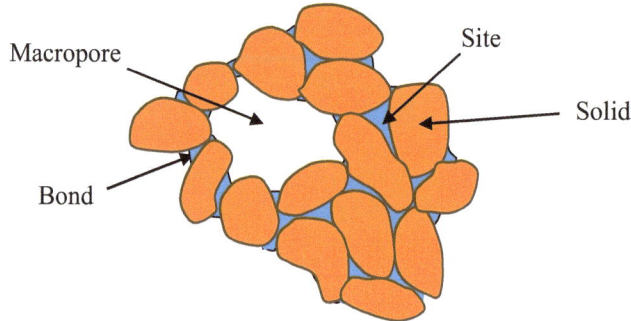

Fig. (2). Soil structure: macropores, mesopores, bonds and solids.

Now, consider Eq. 2.15 (Chapter 2) representing Bishop's parameter χ. This relationship can be written in the following form

$$\chi = f^s + S_w^u f^u = \sum_f S_w^i f^i$$

where superindex i represents the type of fraction: dry (d), saturated (s) or unsaturated (u). In terms of the volumetric behavior the above equation can be interpreted as follows: it represents the addition of the product of every fraction by its degree of saturation. This means that for the saturated fraction, the effect of an increment of suction is the same as an increment of the mean net stress (term s f^s in Eq. (2.14), Chapter 2) because in that case $S_w^s = 1$, whereas for the unsaturated fraction, the effect of an increment of suction is proportional to the degree of saturation of this fraction (term $s\, S_w^u\, f^u$ in Eq. (2.14)). Finally, the dry fraction (f^d) shows a nil degree of saturation and therefore does not appear in Eq. (2.14). In other words, Bishop's parameter represents the weighted degree of saturation of all three fractions.

Even if the dry fraction does not play any role in the volumetric behavior of the soil during suction increase nor appears in the determination of parameter χ, it certainly plays a role during mean net stress increase as in this case all fractions participate equally on the volumetric deformation of the soil. This means that the term $\bar{p}$ in Eq. (2.14) is multiplied by the sum of all fractions which is equal to one. This agrees with the description of the experimental behavior of unsaturated soils made before.

It is noteworthy that Eq. (8.4) uses the same compressive index λ_e no matter if it corresponds to an increase in mean net stress or suction although, for this last case, the index is affected by parameter χ because it represents the proportion of pores affected by suction changes. In other words Eq. (8.4) represents a single compression curve for both net stress and suction increase.

8.3. ELASTOPLASTIC FRAMEWORK

In addition to Eq. (8.4), the modeling of the volumetric behavior of unsaturated soils requires an elastoplastic framework. The framework considered herein is sketched in Fig. (3) in the axes of mean net stress (Fig. 3a) and mean effective stress (Fig. 3b). A normally consolidated soil sample exhibits a loading collapse yield surface (LCYS) represented by a line forming an angle of 135° with the mean net stress axis as established by Sheng, Sloan and Gens [25] (Fig 3a). When this sample is subjected to a suction s, the drying path (represented by a vertical line in Fig. 3a) crosses the initial yield surface generating a plastic deformation. This plastic deformation produces the hardening of the LCYS which displaces to the right hand side. This displacement depends on the increment of the effective stress applied to the soil and is represented by the value of the matric suction stress χs. In other words, χs represents the increment of the mean net stress that produces the same volumetric plastic deformation generated during the drying of the soil. If at this point the soil is wetted up to saturation, it follows an elastic unloading that does not affect the position of the LCYS. Thus, it can be inferred that the preconsolidation stress in saturated conditions has also increased in the quantity χs. This means that the LCYS can be represented by a vertical line in the mean net stress plane and therefore, it shows the same shape of the drying path

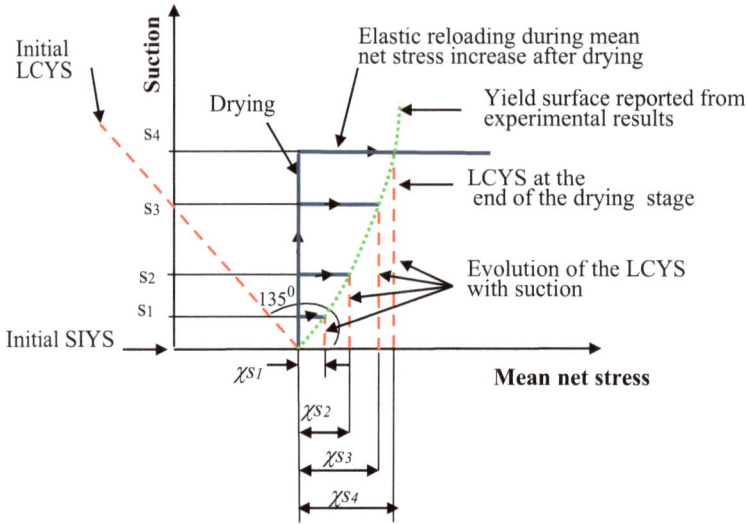

(a)

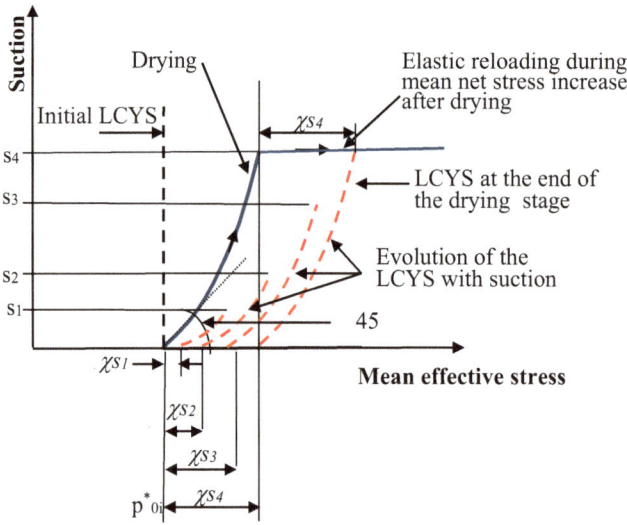

(b)

Fig. (3). Evolution of the LCYSs during drying in (**a**) the mean net stress axis and (**b**) mean effective stress axis.

but displaced in the quantity χs. If the intersecting points of this surface with the loadings paths followed by net stress increase at constant suction are linked together by a line (fine dotted line in 8.3(a)), it adopts the shape of the LCYS that has been experimentally determined by different researchers (see for example [92,

95, 108]). It can be argued (as Sheng [94] does) that the experimental procedure used to obtain the LCYS considers that the soil initially behaves elastically during the loading stage after drying but this could not be the case. Only the analysis of more experimental results would give light to this issue.

In the axis of the mean effective stress, the drying path initially adopts a slope of 45° and then deviates from this direction as the soil becomes unsaturated. Similarly, the LCYS adopts this same shape and displaces the quantity χs on the mean effective stress axis as shown in Fig. (**3b**).

According to this framework, when Eq. (8.4) is plotted for a set of tests performed at constant suction, each sample suffers a different hardening given by the quantity χs and therefore, the loading paths are represented by a family of curves in the axes of the logarithm of the mean net stress *versus* void ratio as shown in Fig. (**4a**). These curves are similar to the volumetric response of unsaturated soils reported by different researchers (see for example: [83, 92, 107, 109]). When these results are plotted in the axes of the logarithm of the mean effective stress versus void ratio, a family of curves, as those shown in Fig. (**4b**), is obtained. In this case, the data reported by Futai and Almeida [92] for a particular soil was used in order to establish the values of Bishop's parameter χ at different suctions. This procedure is shown in the next section of this chapter. In general, these curves can be assimilated to parallel straight lines for small ranges of the mean effective stress. Because parameter χ depends on the SWRCs of the material, the amount of suction hardening also depends on these curves. In general, soils showing large ranges of suction in their SWRCs exhibit large suction hardening. On the contrary, suction hardening is difficult to observe in soils showing a small range of suction in their SWRCs.

Elastic behavior of the material occurs when the current mean effective stress is smaller than the maximum mean effective stress experienced by the soil. For such a case, parameter λ_e becomes the slope of the unloading-reloading stress path κ_e, and relationship (8.4) transforms into

$$\frac{e}{e_0} = \left(\frac{p_0'}{p'}\right)^{\kappa_e} \tag{8.5}$$

This behavior happens when the mean net stress applied to the soil reduces while suction remains constant but, this may eventually happen when the soil attains large suctions because the matric suction stress reduces after reaching a maximum as shown in the previous chapter.

(a)

(b)

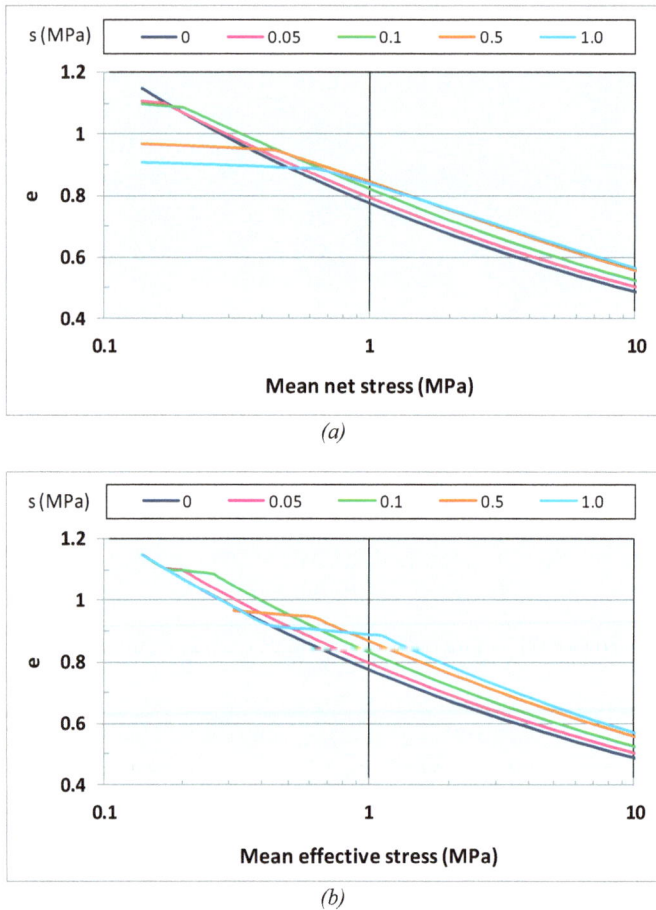

Fig. (4). Numerical volumetric behavior of soils related to (**a**) mean net stress and (**b**) mean effective stress.

8.4. NUMERICAL AND EXPERIMENTAL COMPARISONS

In order to evaluate the proposed framework for the volumetric behavior of unsaturated soils, the results of various tests made on a variety of materials and for different loading conditions were employed. Fleureau, Kheirbek-Saoud, Soemitro and Taibi [93] prepared different clayey soils at a water content of 1.5 times its liquid limit. This slurry was consolidated in an oedometer cell with vertical stresses ranging between 0.06 and 0.2 MPa. Then the samples followed a drying-wetting path where suction was controlled using the axis translation technique for low suctions and the vapor circulation technique for large suctions. With the results of these tests, the variations of the void ratio, the degree of saturation and the water content with the value of suction were determined. Also the GSD and both SWRCs for some of these materials were reported. Here the results of the Sterrebeek loam, Montmorillonite clay and Yellow clay are presented. As the PSD of the soil was not available, it was inferred by fitting the numerical SWRCs with the experimental results according to the procedure outlined in Chapter 5.

The numerical PSD along with the experimental GSD and the void ratio were used to build up the porous-solid model from which the value of parameters f^s, f^u and S_w^s were determined as a function of suction for both the wetting and the drying path. This allowed determining the mean effective stress given by Eq. (2.14) (Chapter 2) and the numerical volumetric response of the material was derived from Eq. (8.4). Fig. (**5a**) shows the fitting of the SWRCs for the Montmorillonite clay. Fig. (**5b**) represents the numerical PSD obtained at the end of the fitting process along with the experimental GSD of the material in the axes of size *versus* relative volume. The relative volume is the volume of pores (or solids) of certain size divided by the total volume of pores (or solids) as explained before. Notice the similarity in shape of these two curves as it has already been pointed out by Alonso, Rojas and Pinyol [52]. Fig. (**5c**) shows the values of parameters f^s, f^d, f^u, S_w^s and χ obtained from the porous-solid model during a drying path. It can be observed that the saturated fraction starts reducing at very low suctions while the dry fraction only appears at very large suctions meanwhile the unsaturated fraction increases up to a certain point and then decreases to become nil at very large suctions. It can also be observed that parameter χ is closely related to the saturated fraction of the soil. On the other hand, the degree of saturation of the unsaturated fraction S_w^u shows important fluctuations at low values of suction. This happens because most of the soil is still saturated at these values of suction and the unsaturated fraction (represented by those solids and their surrounding pores showing a combination of saturated and dry pores) is constituted by a small number of elements (solids, cavities and bonds). Therefore a small change in the number of saturated elements produces important changes on the value of this parameter.

Only when the number of elements of this fraction is large enough it shows a continuous increase and, at certain point it reduces to become nil at large suctions. Finally, the numerical prediction and the experimental results of the volumetric behavior upon drying are compared in Fig. (**5d**). For each increment of suction the value of parameters f^s, f^u and S_w^u are obtained, the value of parameter χ is computed and the new void ratio is calculated. These results were obtained for $\lambda_e = -0.36$.

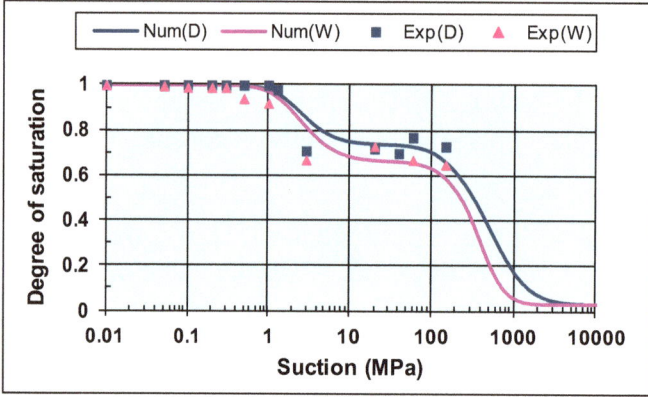

(a)

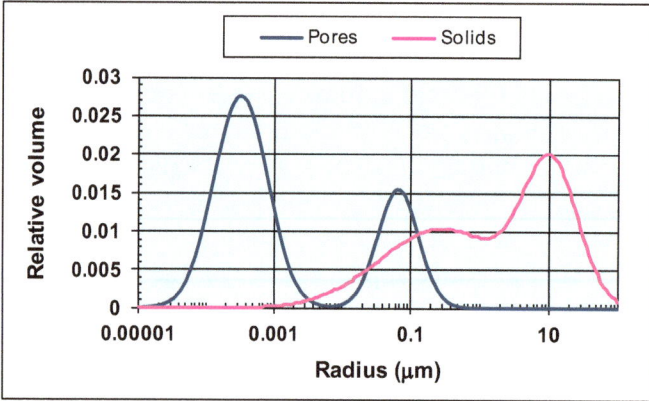

(b)

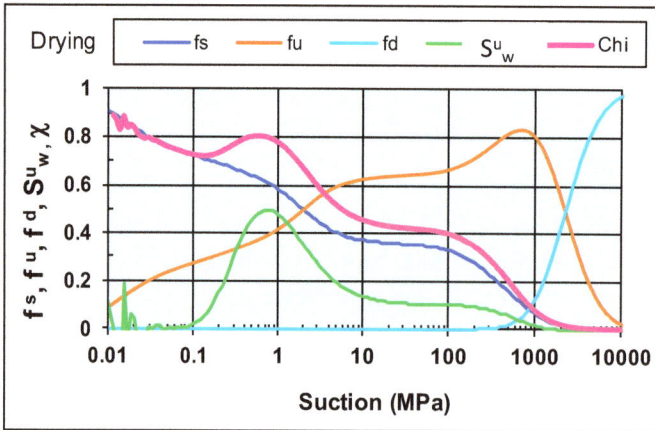

(c)

Fig. 5 cont.....

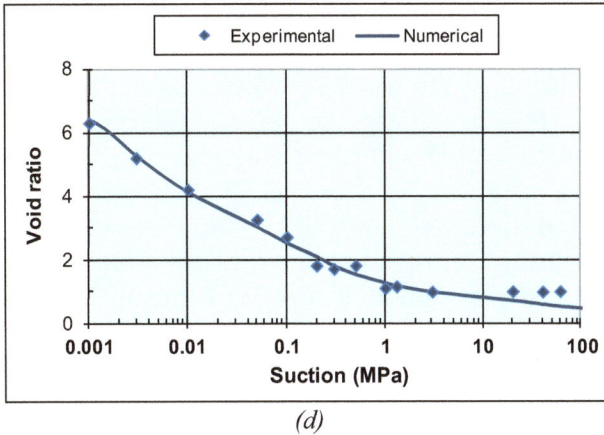

(d)

Fig. (5). Results for the Montmorillonite clay: (**a**) fitting of the SWRCs, (**b**) numerical PSD and experimental GSD, (**c**) parameters f^s, f^u, S_w^S and χ and (**d**) volumetric behavior by suction increase (experimental data from [93]).

Following the same procedure the results of the Sterrebeek loam were simulated. Fig. (**6a**) shows the fitting of the wetting (W) and drying (D) SWRCs. Fig. (**6b**) shows the values for parameters f^s, f^u, f^d and S_w^S at drying. These same parameters can be found at wetting. With these values parameter χ can be obtained for both conditions: wetting and drying as shown in Fig. (**6c**). Finally, Fig. (**6d**) presents the comparison between numerical and experimental results for the volumetric behavior of this material. In this case, the volumetric response of the slurry following a drying path was determined for two different initial conditions: at zero net stress and at a preconsolidated net stress of 0.2 MPa. These results were obtained with $\lambda_e = -0.12$. For both cases, it can be said that Eq. 8.4 is adequate for simulating the volumetric behavior of the material.

Fig. (**7**) shows the results obtained for the Yellow clay. Fig. (**7a**) shows the fitting of the SWRCs at wetting (W) and drying (D). Fig. (**7b**) shows the values for parameters f^s, f^u, f^d and S_w^S at drying. The values for parameter χ at wetting and drying are shown in Fig. (**7c**). Finally, Fig. (**7d**) presents the comparison between numerical and experimental results for the volumetric behavior of this material following wetting and drying paths. These results were obtained for $\lambda_e = -0.13$ and $\kappa_e = -0.013$. It can be observed that for both stress paths the model adequately simulates the behavior of the material.

Futai and Almeida [92] reported the results of different tests made on undisturbed samples of a residual soil. Additional data included the GSD, the PSD obtained from MIP tests and both SWRCs obtained by combining the suction plate and the filter paper technique. Fig. (**8**) shows the results of isotropic compression tests made on samples subjected to three different suctions.

Fig. (**8a**) shows the fitting of the SWRCs which resulted in the numerical PSD shown in Fig. (**8b**). In this last figure, the experimental results of a MIP test are also included and compare well with the numerical results. Fig. (**8c**) shows the values of parameter χ obtained from Eq. (2.15), Chapter 2. Finally, Fig. (**8d**) shows the comparison between numerical and experimental results of the volumetric behavior during isotropic loading tests on samples subject to different suctions. It can be noted that although there is some scattering for the sample tested at a suction of 0.3 MPa, the proposed framework is able to adequately simulate the volumetric behavior of soils during this type of test.

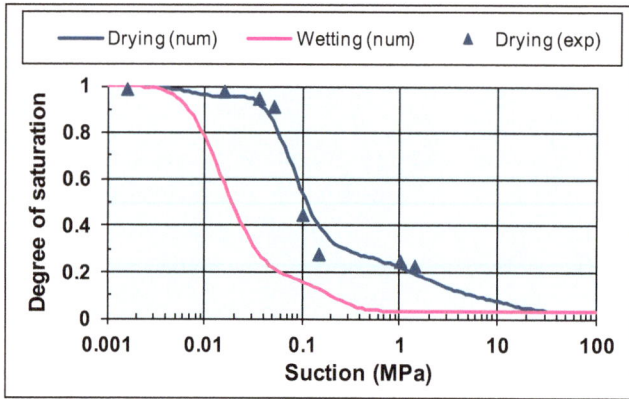

(a)

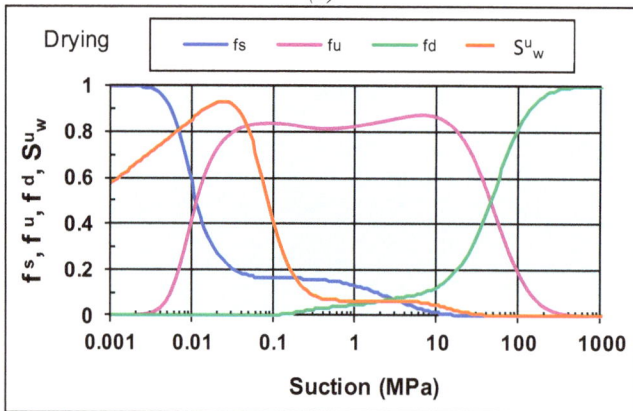

(b)

Fig. 6 cont.....

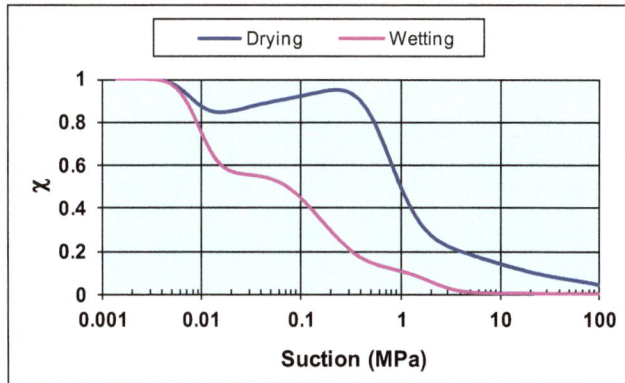

(c)

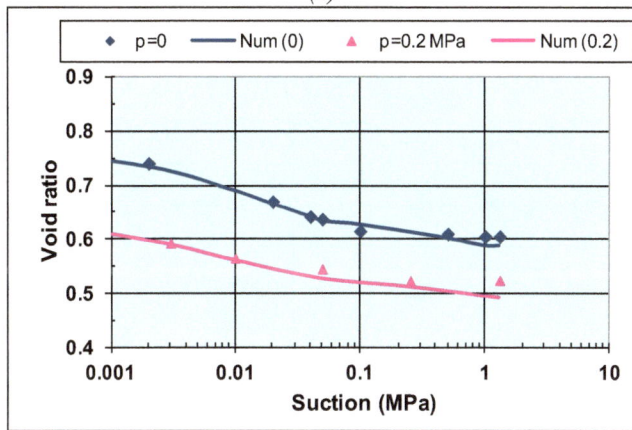

(d)

Fig. (6). Results for Sterrebeek loam (**a**) fitting of the SWRCs, (**b**) parameters f^s, f^u and S^S_W, (**c**) parameter χ and (d) volumetric behavior by suction increase at two different net stresses (experimental data from [93]).

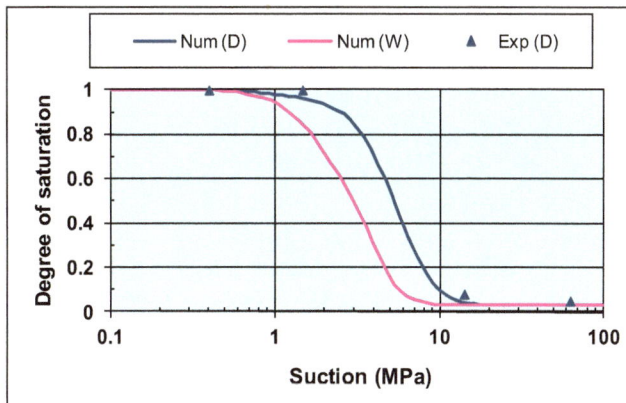

(a)

Fig. 7 cont.....

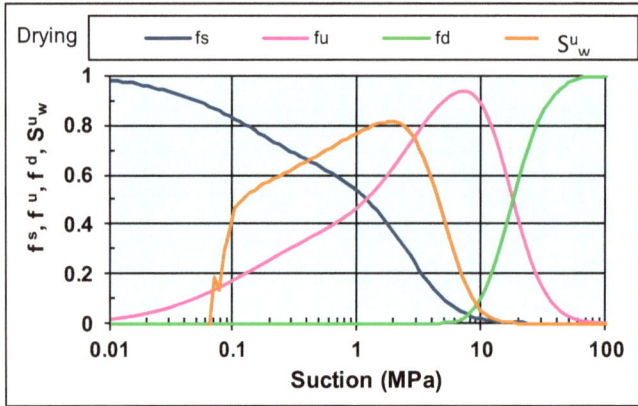

(b)

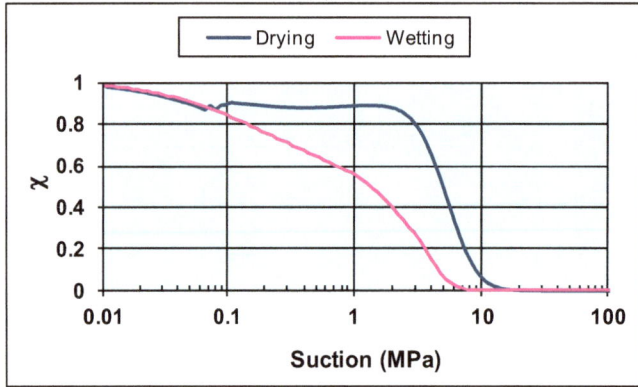

(c)

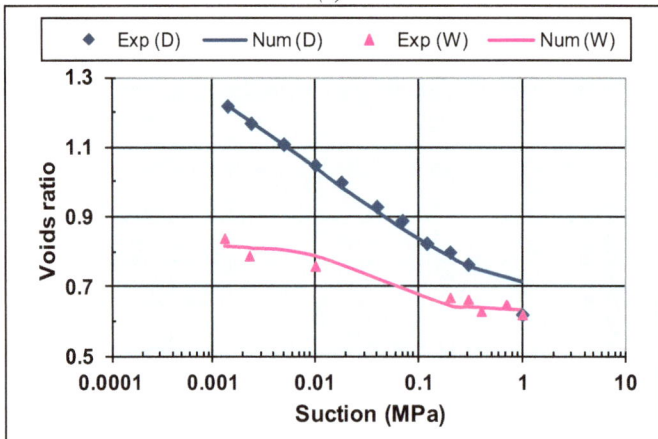

(d)

Fig. (7). Results for Yellow clay (**a**) fitting of the SWRCs, (**b**) parameters f^s, f^u and S_w^u (**c**) parameter χ and (**d**) volumetric behavior by suction increase at two different net stresses (experimental data from [93]).

The volumetric strains were determined using Eq. (8.5) up to the yield stress generated at the end of the drying stage. From that point, Eq. (8.4) was used for the rest of the curve. Because the increase in the yield stress of a saturated sample generated during a drying stage is given by the quantity χs, the final yield stress is obtained by adding the initial preconsolidation stress of the soil in saturated conditions to the increment of the yield stress during the drying stage. For a small increment of suction (ds), the increase in the yield stress is given by $d\bar{p} = \chi \, ds + s \, d\chi$ where ds represents the increment of suction and $d\chi$ the variation of Bishop's parameter due to that increment. The preconsolidation stress of the material in saturated conditions is around 0.14 MPa as it can be observed in Fig. (**8d**). The values of parameter χ for the different suctions were obtained from Fig. (**8c**) resulting in: $\chi = 0.6$ for $s = 0.1$ MPa and $\chi = 0.58$ for $s = 0.3$ MPa which produced the following preconsolidation stresses: $\bar{p}_0 = 0.2$ for s = 0.1 MPa and $\bar{p}_0 = 0.31$ for s = 0.3 MPa. From the results of Fig. (**8d**) it can be noticed that the theoretical preconsolidation stress for the unsaturated samples corresponds well with the experimental results. The considered values for the compression index at loading and reloading were $\lambda_e = -0.25$ and $\kappa_e = -0.04$.

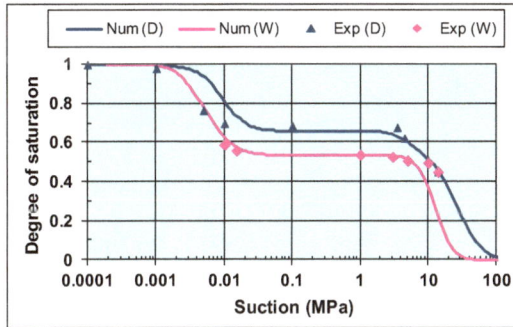

(a)

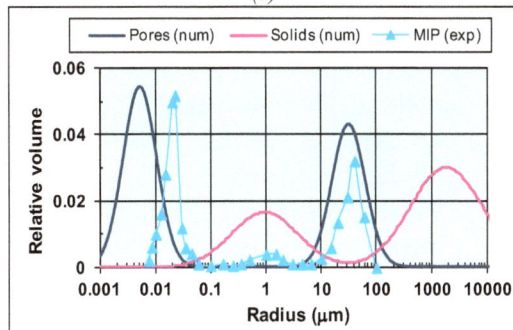

(b)

Fig. 8 cont.....

(c)

(d)

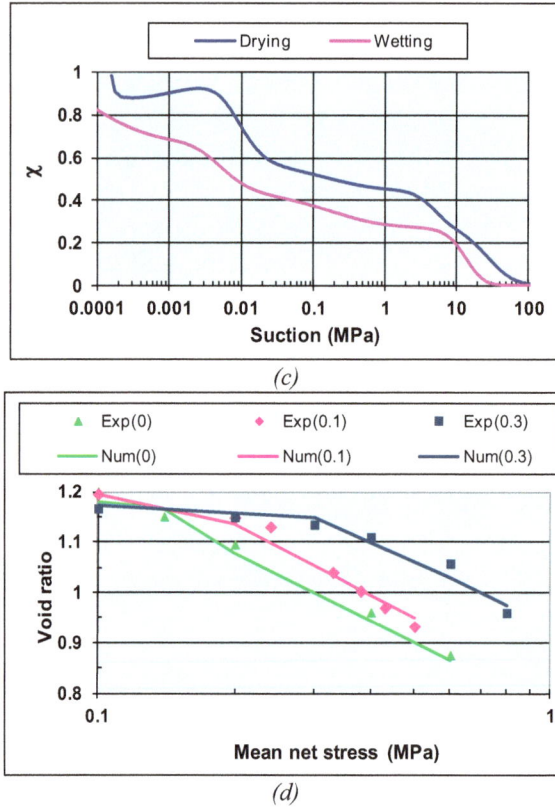

Fig. (8). Results for a residual gneiss (**a**) fitting of the SWRCs, (**b**) numerical and experimental PSD, (**c**) values of parameter χ and (**d**) volumetric behavior at different suctions (experimental data from [92]).

Similar tests were performed by Cunningham, Ridley, Dineen and Burland [110] in a mixture of 20% speswhite kaolin, 10% of London clay and 70% of silica silt. With this mixture, a slurry was prepared at a water content 1.5 times its liquid limit. Then it was pre-consolidated one-dimensionally in a 20.4 cm diameter oedometer to a maximum vertical stress of 0.2 MPa. The soil samples were then trimmed from this pre-consolidated soil mass to the appropriate size. The tests included the GSD, both SWRCs obtained from the filter paper technique, isotropic loading and shearing tests at constant suction in the triaxial apparatus. For these last tests, suction was controlled using the air circulation technique and was measured directly in the soil samples using two suction probes. Some of the experimental results are shown in Fig. (**9**). The experimental SWRCs and the numerical fitting for these curves are shown in Fig. (**9a**). Fig. (**9b**) shows the variation of parameter χ with suction. With these values it is possible to obtain the numerical volumetric response of the material during suction increase as shown in

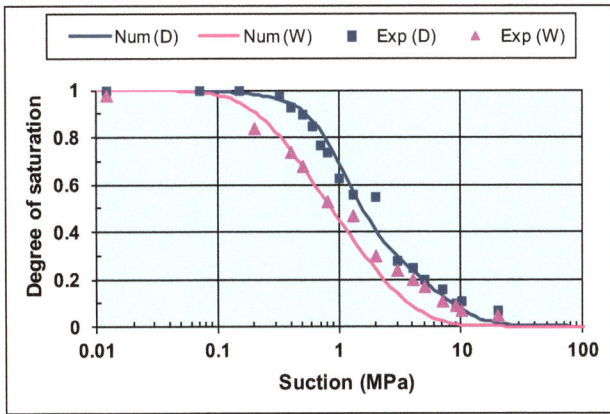

(a)

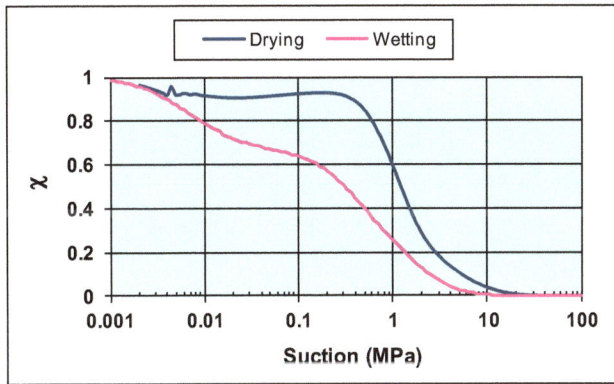

(b)

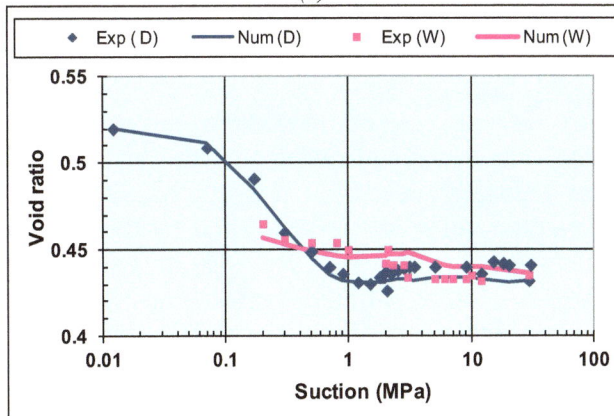

(c)

Fig. 9 cont.....

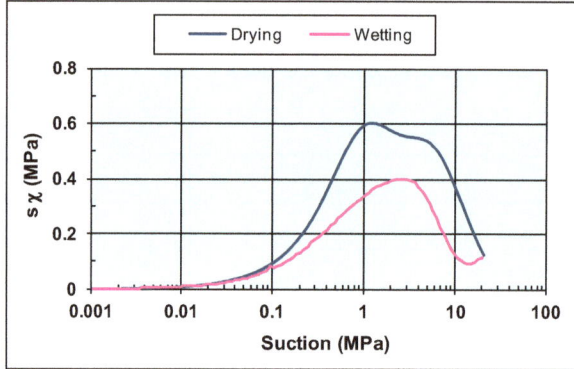

(d)

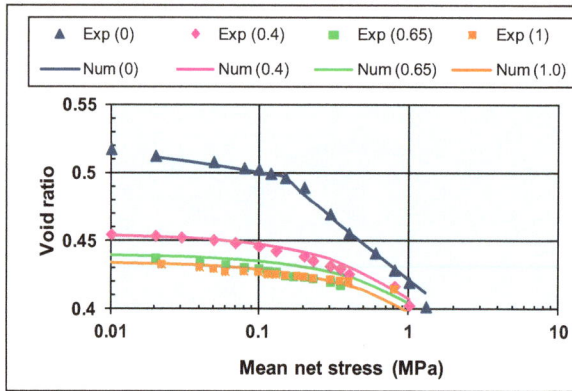

(e)

Fig. (9). Results for a soil mixture: (**a**) fitting of the SWRCs, (**b**) parameter χ, (**c**) volumetric response with suction, (**d**) matric stress $s\chi$ and (**e**) volumetric response with mean net stress (experimental data from [110]).

Fig. (**9c**). In this case, the experimental behavior shows a clear elastic rebound (even if there is some scattering) indicating that the effective stress reduces at some stage during the drying process. This happens because as suction increases, the value of the matric suction stress (represented by the product $s\chi$) reaches a maximum and then decreases, as shown in Fig. (**9d**). In this figure, it can be noted that when suction reaches a value slightly greater than 1 MPa, the matric suction stress reaches its maximum and then reduces while suction keeps increasing. When the drying path inverses to wetting, matric suction stress reduces further then increases (but never reaches the drying maximum value) and finally reduces again while suction keeps reducing. Therefore, when the matric suction stress reduces after reaching its maximum value, the numerical response switches from elastoplastic (Eq. (8.4)) to purely elastic (Eq. (8.5)). The numerical results were obtained with the following parameters $\lambda_e = -0.13$ and $\kappa_e = -0.04$. The switch from

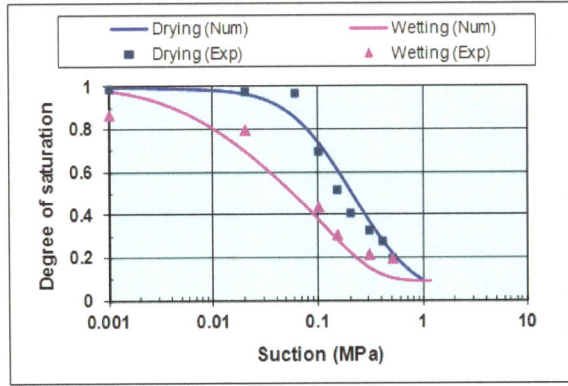

(a)

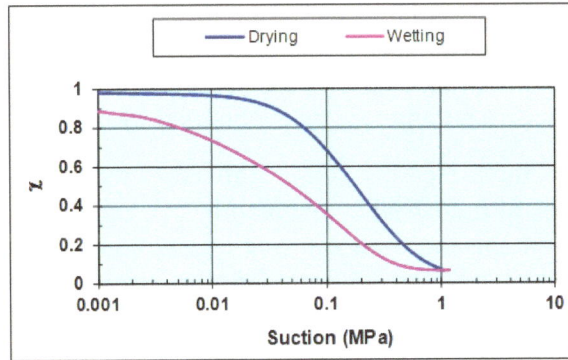

(b)

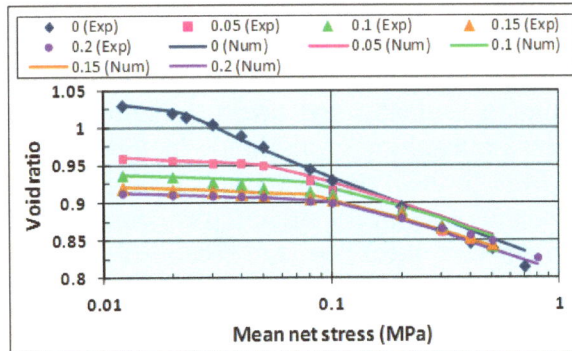

(c)

Fig. (10). Results for a coarse kaolin: (**a**) fitting of the SWRCs, (**b**) parameter χ in wetting and drying and (**c**) volumetric behavior (experimental data from [95]).

elastoplastic to elastic behavior during suction increase has been well documented by Khalili, Geiser and Blight [111]. Similar results were reported and simulated by Vlahinic, Jennings and Thomas [112] and Blight [113] on different porous materials. This reduction in the effective stress during drying can also be observed on the experimental results reporting the strength of the material with suction. A maximum value and then a reduction of the strength can be observed if suction is sufficiently increased (see for example: [21, 34, 35, 70, 112, 114, 115]). This same behavior is observed in the tensile strength of soils as shown in the previous chapter and also reported by Fredlund, Xing, Fredlund and Barbour [71]. Also some constitutive models consider this reduction on the strength of soils with suction (see for example [70, 116, 117]). Not all soils show this behavior, for example soils with large clay contents may show a continuous increase of strength with suction due to the presence of menisci of adsorbed water that do not disappear even at very large suctions. On the contrary, the effect of suction on the strength of sandy soils completely disappears at large suctions. Fig. (**9e**) shows the numerical and experimental results for isotropic loading at different suctions on the axis of the mean net stress *vs.* void ratio. In this case the materials were dried from slurry and because of that, the preconsolidation stress equals the matric suction stress $s\chi$, where parameter χ was obtained from Fig. (**9b**).

Other tests were conducted by Thu, Rahardjo and Leong [95] on statically compacted samples of industrial coarse kaolin. All samples were compacted at the optimum water content and then saturated using back pressure. Afterwards, all samples were consolidated at an isotropic net stress of 0.01 MPa following a drying stage where suction ranged between 0 and 0.3 MPa. Finally, the samples were isotropically loaded up to 0.7 MPa in net stress. The fitting of the SWRC in wetting and drying is shown in Fig. (**10a**). The values of parameter χ in wetting and drying are shown in Fig. (**10b**). Finally, Fig. (**10c**) shows the comparison between experimental and numerical results for the isotropic loading tests performed at different suctions. The preconsolidation stress for each test was obtained by adding to the saturated preconsolidation stress (0.025 MPa), the matric suction stress χs, resulting in the following preconsolidation stresses: 0.07, 0.09, 0.1 and 0.1 MPa for suctions of 0.05, 0.1, 0.15 and 0.2 MPa, respectively. The values of parameter χ for every suction were obtained from Fig. (**10b**). These results show good agreement for both the preconsolidation stress and the overall volumetric behavior of the material.

CHAPTER 9

Collapse Upon Wetting

Abstract: This chapter presents the modeling of the phenomenon of collapse upon wetting using the effective stress equation established in Chapter 2 and the elastoplastic framework proposed in the previous chapter. Using the probabilistic porous-solid model, Bishop's parameter χ can be obtained to determine the current effective stress. The proposed framework includes the hysteresis of the SWRC and to some extent the hydro-mechanical coupling of unsaturated soils. This model is able to reproduce some particularities of the phenomenon of collapse upon wetting that other models cannot simulate.

Keywords: Bishop's effective stress equation, Collapse upon wetting, Compacted soils, Effective stress, Elastoplastic framework, Hydro-mechanical coupling, Hysteresis, Neutral loading, Porous-solid model, Preconsolidation stress, Soil-water retention curve, Suction controlled tests, Suction hardening, Unsaturated soils, Yield surface.

9.1. INTRODUCTION

The Barcelona Basic Model (BBM) [7] has been able to reproduce the main aspects of the phenomenon of collapse upon wetting using the independent stress variables principle (Fredlund and Morgenstern, 1977). The key point for the simulation of collapse in this model is the consideration that the apparent preconsolidation stress increases with suction (Fig. **1**). This feature is introduced into the model through the loading collapse yield surface (LCYS) which adopts the geometry shown in Fig. (**1b**). By analyzing the volumetric behavior of a soil sample subject to a drying-wetting cycle, the equation relating the yield stress in unsaturated($\bar{p}_0$) and saturated $(\bar{p}_0^*)$ conditions is obtained as a function of the slopes of the loading ($\lambda(0)$ and $\lambda(s)$) and unloading-reloading (κ and κ_s) curves of the soil in saturated and unsaturated conditions, respectively. This equation writes

$$\frac{\bar{p}_0}{p^r} = \left(\frac{\bar{p}_0^*}{p^r}\right)^{\frac{\lambda(0)-\kappa}{\lambda(s)-\kappa_s}}$$

where p' represents a reference pressure. In general, it can be considered that $\kappa = \kappa_s$ as their values are relatively small. In turn, $\lambda(s)$ depends on the values of $\lambda(0)$ and suction. This equation represents the shape of the LCYS as shown in Fig. (**1**). When an increment of the net stress is applied to an initially saturated sample that has been dried to suction s (stress path AA'BD in Fig. **1**), the initial $LCYS_i$ displaces on the mean net stress axis reaching the position $LCYS_f$.

The volumetric compression of the material during a net stress increase (dv_p) beyond the yield stress is given by

$$dv_p = -\lambda(s)\frac{dp}{p}$$

In the same way, the volumetric response of the soil during a suction increase (dv_s) beyond the suction increase yield surface (SIYS) is given by

$$dv_s = \lambda_s \frac{ds}{s+p_{at}}$$

where λ_s represents the slope of the virgin compression line (VCL) during suction increase, ds is the increment in suction and p_{at} is the atmospheric pressure. In the case when both net stress and suction are increased, the total volumetric response of the material (dv) is given by the addition of the volumetric behavior during mean net stress increase and suction increase, in the form

$$dv = dv_p + dv_s$$

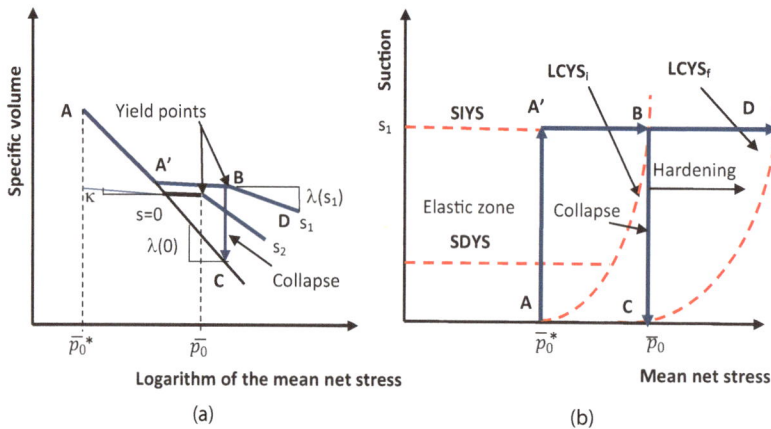

Fig. (1). (a) Volumetric behavior and **(b)** hardening of the LCYS in the BBM.

This model has been widely employed to reproduce the volumetric behavior of unsaturated soils with great success.

Other models include the effective stress in their formulations for example Bolzon, Schrefler and Zienkiewics [118], Vaunat, Romero and Jommi [90], Loret and Khalili [97], Karube and Kawai [119], Gallipoli, Gens, Sharma and Vaunat [23], Wheeler, Sharma and Buisson [22]. These models make use of Bishop's [14] effective stress σ'_{ij}

$$\sigma'_{ij} = \sigma_{ij} - \delta_{ij}[u_a - \chi(u_a - u_w)]$$

The Bishop parameter χ can be written as a function of the degree of saturation, suction or both. For example the model proposed by Gallipoli, Gens, Sharma and Vaunat [23] considers that χ is equal to the degree of saturation. To account for the volumetric behavior of unsaturated soils, the model includes the constitutive parameter ξ which represents the bonding and debonding stress produced by water menisci. This parameter is written as a function of the degree of saturation S_r in the form

$$\xi = f(s)(1 - S_r)$$

where *f(s)* is a function of suction representing interparticle forces. By analyzing the relationship between the ratio e/e_s with parameter ξ, where e and e_s represent the voids ratio in unsaturated and saturated conditions when the soil is subject to the same Bishop stress, the flowing relationship was established

$$\frac{e}{e_s} = 1 - a[1 - exp(b\xi)]$$

With this model, it is possible to model the volumetric response of soils subjected to increments of the mean net stress and wetting-drying cycles including the phenomenon of collapse.

Another variation for the value of parameter χ is presented by Alonso, Pereira, Vaunat and Olivella [120]. In this case, the global degree of saturation (S_w) of the material is split in two parts: one for the macrostructure S_w^M and another for the microstructure S_w^m in the form $S_w = S_w^M + S_w^m$. Then two possible equations for the value of parameter χ are proposed. The first one is given by the expression

$$\chi \equiv S_w^e = \langle \frac{S_w - S_w^m}{1 - S_w^m} \rangle$$

where S_w^e is the effective degree of saturation and $\langle x \rangle = (x + |x|)/2$ represents the Macauley brackets. The other possibility is given by the expression

$$\chi = (S_w)^a$$

where *a* represents a soil parameter larger or equal to 1. With this model, the authors have been able to reproduce the strength and volumetric behavior of unsaturated soils.

The model by Della Vecchia, Jommi and Romero [121] also uses the concept of micro and macroporosity. It includes the evolution of the SWRC related to the volumetric deformation of the material and the hysteresis of the SWRC. This model is able to reproduce a maximum collapse strain with increasing confining stress.

The model proposed by Wheeler, Sharma and Buisson [22] considers a vertical LCYS in the plane of modified suction against mean Bishop's stress. This surface is coupled to the SI and the suction decrease yield surfaces (SDYS) which are represented by horizontal lines in the same plane. Both the SIYS and SDYS surfaces are coupled with the LCYS. For example when the soil dries and suction exceeds the maximum value experienced by the soil, plastic strains occur that hardens the SIYS. This hardening is represented by a vertical displacement of the surface that pulls along the SDYS, while the LCYS moves outwards. On the contrary, when suction reduces beyond the SDYS this surface is pulled downwards along with the SIYS, while the LCYS moves inwards. This coupling between the yield surfaces allows modeling the phenomenon of collapse upon wetting during the inward movement of the LCYS, the wetting path remains outside the elastic zone.

However, these models show deficiencies in reproducing one or more of the following particularities of the phenomenon of collapse upon wetting observed on compacted materials tested at different densities and loaded at different net stresses: a) In most cases, there is an initial elastic response at the beginning of collapse. The extent of this elastic response depends on the value of the mean net stress applied during the loading stage b) collapse deformation depends on both the density and the stress state before collapse c) For increasing values of the mean net stress applied during the loading stage on samples compacted at the same density, the collapse volumetric strain increases and then decreases (*i.e.* there is a maximum collapse volumetric strain for samples compacted at the same void ratio). This behavior has been experimentally observed by different researchers for example: Sun, Sheng, Cui and Sloan [122]; Rodrigues and Volar [123].

This chapter shows the modeling of the phenomenon of collapse upon wetting based on the principle of effective stress resulting in a model able to reproduce the particularities listed above. This treatment also results in a unifying framework for saturated and unsaturated soils.

9.2. VOLUMETRIC ELASTOPLASTIC FRAMEWORK

Based on the elastoplastic framework developed in the preceding chapter, it is possible to simulate the phenomenon of collapse upon wetting of soils. Consider a saturated normally consolidated sample that has been dried to suction s_0, then loaded by an increase of the mean net stress in the quantity $\chi_0 s_0$ (path BC in Fig. 2). In these conditions, the sample will be unable to collapse as the wetting path (WP) lays inside the elastic zone (see Fig. 2). On the contrary, if the increment of the mean net stress ($\Delta\bar{p}$) is larger than $\chi_0 s_0$ (path BE in Fig. 2), the LCYS tilts further showing smaller slopes (curve EG) and just then, the soil would be able to collapse when wetted.

The tilt of the LCYS occurs because the upper part of the curve follows the mean effective stress increment ($\Delta\bar{p}$) applied during loading (point E) while the lower part hardens to the apparent preconsolidation stress ($\Delta\bar{p}_0$) in saturated conditions (point G). This increment in the apparent preconsolidation stress in saturated conditions can be obtained by computing the plastic volumetric strain generated during the loading stage from points C to E in Fig. (2).

$$\Delta\varepsilon_v^p = \left(\frac{e_E}{1 + e_C}\right)(\lambda_e - \kappa_e)\left(\frac{\Delta\bar{p} - \chi_0 s_0}{p_{0i}' + \chi_0 s_0 + \Delta\bar{p}}\right)$$

where e_C and e_D represent the void ratio at points C and D, respectively. p_{0i}' is the initial mean effective stress applied to the sample in saturated conditions (see Fig. 2). This plastic strain should be the same when an increment of the mean net stress ($\Delta\bar{p}_0$) is applied in saturated conditions (point D). And therefore, the value of $\Delta\bar{p}_0$ can be obtained with the relationship

$$\Delta\bar{p}_0 = \Delta\varepsilon_v^p\left(\frac{1 + e_D}{e_G}\right)\frac{p_{0i}' + \chi_0 s_0 + \Delta\bar{p}_0}{(\lambda_e - \kappa_e)} = \left(\frac{1 + e_D}{1 + e_C}\right)\left(\frac{e_E}{e_G}\right)\left(\frac{p_{0i}' + \chi_0 s_0 + \Delta\bar{p}_0}{p_{0i}' + \chi_0 s_0 + \Delta\bar{p}}\right)(\Delta\bar{p} - \chi_0 s_0) \quad \textbf{(9.1)}$$

where e_D represents the void ratio at point D. If the soil is unloaded from point C to point D following the LCYS after drying (dotted curve CD), then the void ratio at point D will be the same as in point C. Moreover, assuming that the variation in void ratio from point E to G is small then Eq. (9.1) simplifies to

$$\Delta\bar{p}_0 = \left(\frac{p_{0i}' + \chi_0 s_0}{\bar{p}_0 + 2\chi_0 s_0}\right)(\Delta\bar{p} - \chi_0 s_0)$$

From this last equation, it results $\Delta\bar{p}_0 < (\Delta\bar{p} - \chi_0 s_0)$. Hence, the LCYS tilts further enabling the WP (path EFH) to cross it at point F triggering the collapse of the sample. According to Fig. (2) and, depending on the hysteresis of the SWRC and the stress increment $\Delta\bar{p}$, the soil may show an initial elastic response (path EF) before the beginning of collapse.

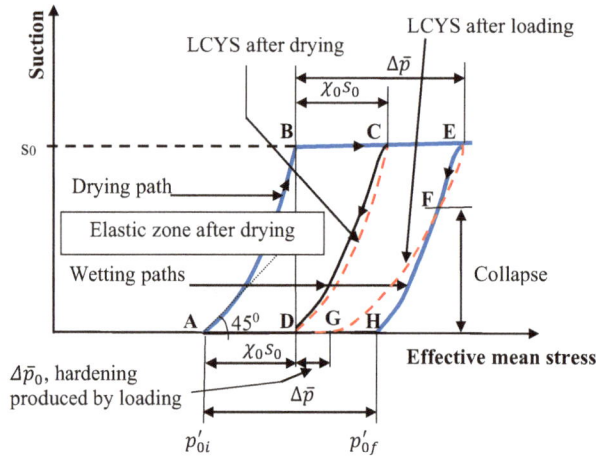

Fig. (2). Collapse upon wetting phenomenon.

Fig. (**2**) shows that the hysteresis of the SWRC can be easily introduced into the elastoplastic framework when plotted in the effective stress plane as the processes of drying and wetting follow different paths. In addition, when effective stresses are used, the hydro-mechanical coupling observed in unsaturated soils is implicit in the formulation. For these reasons, the effective stress representation is adopted herein.

9.3. COMPACTED SOILS

Now, consider that a soil sample is prepared by static compaction in different layers. Because the stress path followed by the soil during compaction cannot be easily determined, it is difficult to establish the initial position of the LCYS even if the applied stresses during compaction and the value of suction in the "as compacted" state are known. Nevertheless, some considerations can be done to establish this initial position. During static compaction, stresses are applied on top of a disaggregated material placed inside a mould. These stresses produce the displacement and interlock of solid particles. The interlock of solid particles is favored by the water menisci located at the points of contact between particles. During loading, the structure of the soil is compressed and then decompressed when unloaded. At the same time, suction decreases during compression and increases during decompression [124]. Then, at the end of compaction the soil lays close to the drying branch of the SWRC corresponding to the void ratio reached at the end of compaction [125]. It is considered here that the process of compaction induces a mean effective stress on the sample that can be decomposed into a mean net stress called here the interlock or fabric stress p_{fab} and the "as compacted" suction s_{00} (see Fig. **3**). The interlock or fabric stress represents the mean preconsolidation stress of the compacted sample taken by water menisci during the unloading process of compaction and can be determined experimentally in the same way as a preconsolidation stress. A more detail explanation on the fabric stress is given in Chapter 12. It is also considered here that the "as compacted" LCYS initiates at the fabric stress (point A in Fig. **3**) and reaches the point representing the "as compacted" state of stresses of the soil sample (point B in the same figure) following a drying path.

After compaction the soil may be subjected to an equalization stage where a suction s_0 and mean net stress $\bar{p}_0$ are applied to the sample following the stress path BCF shown in Fig. (**3**). When a larger suction (s_0) than that shown by the soil after compaction (s_{00}) is applied to the sample, the LCYS displaces to the position given by curve DE in Fig. (**3**). Afterwards, when the soil is loaded to the net stress $\bar{p}_0$, the yield surface hardens by displacing and tilting towards the right hand side reaching the position shown by curve FG. If at this point the soil is wetted, the phenomenon of collapse may occur only if $\bar{p}_0 > \chi_0 s_0 - \chi_{00} s_{00}$ where $\chi_0 s_0 - \chi_{00} s_{00}$ represents the hardening of the LCYS due to suction increase during the equalization process. If, from point F, the soil is further loaded by applying a mean net stress increment $\Delta \bar{p}_0$ (path FH) then the LCYS displaces and tilts further reaching the position HJ indicated in Fig. (**3**). The suction at which the phenomenon of collapse initiates (point I) depends on the applied suction s_0, the total increment of the mean net stress ($\bar{p}_0 + \Delta \bar{p}_0$) and the SWRCs in the form of parameter χ for wetting and drying paths.

The shape of the hardened LCYS can be determined in a similar way as the usual experimental procedure: a load is applied to produce the hardening of the surface, then following an unloading path a new state of stresses is reached and finally the sample is reloaded until it yields again. In the numerical case it must be ensured that the same plastic volumetric strain is retrieved from a different combination of stresses. These different combinations of the state of stresses need to depart from the previous yield surface. In this case, in order to obtain the equation of the hardened LCYS, points D and S in Fig. (**3**) are considered. These two points are located on the LCYS generated after drying the sample to suction s_0 (dotted line DE). Let e_D and e_s be the void ratios at points D and S, respectively. By following an unloading path coinciding with the LCYS after drying (dotted curve DE), the void ratio at point S will be the same as in point D (*i.e.* $e_s = e_D$). Moreover, the stress increment from point D to point H is $\Delta p' = \bar{p}_0 - \chi_0 s_0 + \chi_{00} s_{00} + \Delta \bar{p}_0$ and the plastic decrement of the void ratio is then

$$\Delta e^p = e_H(\lambda_e - \kappa_e)\left(\frac{\bar{p}_0 - \chi_0 s_0 + \chi_{00} s_{00} + \Delta \bar{p}_0}{p_{fab} + \chi_0 s_0 + \bar{p}_0 + \Delta \bar{p}_0}\right)$$

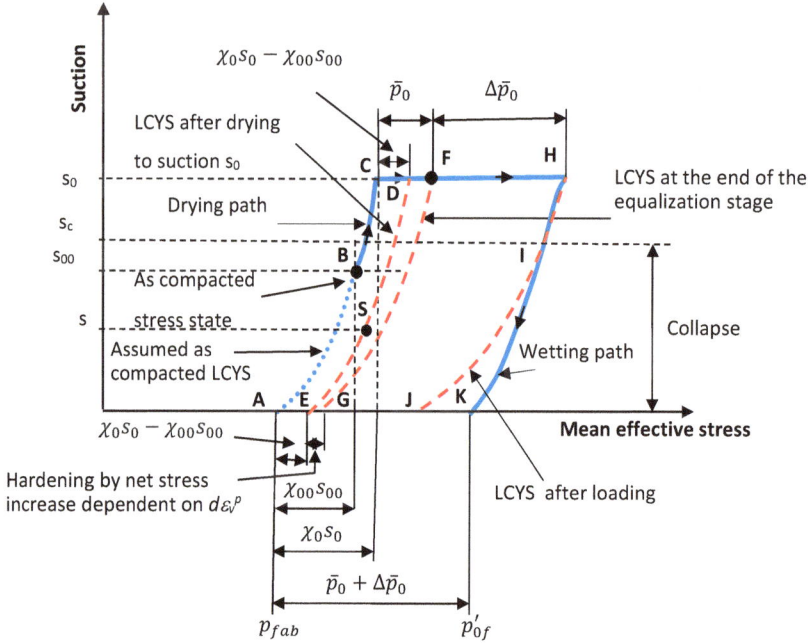

Fig. (3). Elastoplastic framework for the volumetric behavior of compacted soils.

In the same way, the plastic decrement of the void ratio generated by a net stress increase $\Delta \bar{p}_s$ applied at point S is

$$\Delta e^p = e_s(\lambda_e - \kappa_e)\left(\frac{\Delta \bar{p}_s}{p_{fab} + \chi\, s + \chi_0 s_0 - \chi_{00} s_{00}}\right)$$

By equalizing the last two equations the value of $\Delta \bar{p}_s$ can be obtained

$$\Delta \bar{p}_s = \left(\frac{p_{fab} + \chi s + \chi_0 s_0 - \chi_{00} s_{00}}{p_{fab} + \chi_0 s_0 + \bar{p}_0 + \Delta \bar{p}_0}\right)(\bar{p}_0 - \chi_0 s_0 + \chi_{00} s_{00} + \Delta \bar{p}_0) \qquad (9.2)$$

By adopting different values of suction (s), the net stress increment $\Delta \bar{p}_s$ can be obtained and the effective stresses defining the LCYS for different values of suction can be plotted according to the following equation

$$p'_{LCYS} = p_{fab} + \chi s + \chi_0 s_0 - \chi_{00} s_{00} + \Delta \bar{p}_s \qquad (9.3)$$

This means that the LCYS represents a set of equivalent state of stresses able to generate the same plastic volumetric strain for different combinations of net stress and suction. Therefore, when computing the volumetric behavior of soils during wetting, the initial effective stress is represented by a point in the LCYS for the corresponding suction while its increment is obtained from two points on the wetting path representing the change in suction.

Moreover, the equation of the wetting path (WP) initiating at point H (Fig. **3**) is given by

$$p'_{WP} = p_{fab} + \bar{p}_0 + \Delta\bar{p}_0 + \chi_w s \tag{9.4}$$

where χ_w represents the value of parameter χ at wetting. By solving simultaneously Eqs. (9.3) and (9.4) it is possible to define the value of suction at which these curves intersect. This value is called here the collapse suction. Its value is given by

$$s_c = \frac{\chi_0 s_0 (\bar{p}_0 + \Delta\bar{p}_0 - \chi_0 s_0 + \chi_{00} s_{00})}{(\chi_c - \chi_{wc})(p_{fab} + \chi_0 s_0) + \chi_c(\bar{p}_0 + \Delta\bar{p}_0) - \chi_{wc}(\chi_0 s_0 - \chi_{00} s_{00})}$$

where χ_c and χ_{wc} represent the value of parameter χ at the suction collapse in drying and wetting, respectively. For suctions above this value, the soil behaves elastically whereas for smaller values, elastoplastic behavior occurs. The void ratio reached at the collapse suction is

$$e_l = e_{00} \left(\frac{p_{fab} + \chi_0 s_0}{p_{fab} + 2\chi_0 s_0 - \chi_{00} s_{00}} \right)^{(\lambda_e - \kappa_e)} \left(\frac{p_{fab} + \chi_0 s_0 + \bar{p}_0 + \Delta\bar{p}_0}{p_{fab} + \chi_{00} s_{00}} \right)^{\lambda_e} \left(\frac{p_{fab} + \chi_0 s_0 - \chi_{00} s_{00} + \Delta\bar{p}_{sc} + \chi_{wc} s_c}{p_{fab} + \chi_0 s_0 + p_0 + \Delta\bar{p}_0} \right)^{\kappa_e} \tag{9.5}$$

where $\Delta\bar{p}_{sc}$ is given by Eq. (9.2) for $s = s_c$. Once the collapse suction is known, it is possible to follow the deformation of the sample beyond this point using Eq. (8.4, Chapter 8). As sated before, the initial effective stress at a certain value of suction is obtained from the LCYS (curve IJ in Fig. **3**) while its increment is obtained from the wetting path (curve IK). For example, when computing the change in void ratio from the beginning to the end of collapse (wetting path IK), the initial effective stress is represented by point J while its increment is represented by the difference in effective stresses from points I to K, in Fig. (**3**). Under these considerations, it results that the effective stress for a certain value of suction of a collapsing sample during wetting is larger than its equivalent stress previous to wetting and this is precisely the reason why collapse occurs. Then, the initial effective stress at the end of collapse is given by Eq. (9.3) for $s = 0$, while the increment of the effective stress at this same point can be obtained from Eq. (9.4) for the path going from point I ($s = s_c$) to K ($s = 0$) and is equal to $\chi_{wc} s_c$. With these values it is possible to obtain the void ratio reached at the end of collapse

$$e_{K(BCHK)} = e_l \left[1 + \frac{\chi_{wc} s_c (p_{fab} + 2\chi_0 s_0 - \chi_{00} s_{00})}{(p_{fab} + \chi_0 s_0 - \chi_{00} s_{00})(p_{fab} + \chi_0 s_0 + \bar{p}_0 + \Delta\bar{p}_0)} \right]^{\lambda_e} \tag{9.6}$$

Where e_I is given by Eq. (9.5). Similarly, the void ratio reached by loading the sample in saturated conditions following path AK is

$$e_{K(AK)} = e_{00} \frac{\left(p_{fab} + \bar{p}_0 + \Delta\bar{p}_0\right)^{\lambda_e}}{\left(p_{fab} + \chi_{00}s_{00}\right)^{(\kappa_e)}\left(p_{fab}\right)^{(\lambda_e - \kappa_e)}} \tag{9.7}$$

In this case it has been considered that point A is reached from point B following an unloading path, meaning that $e_A = e_B \left(\frac{p_{fab}}{p_{fab} + \chi_{00}s_{00}}\right)^{\kappa_e}$. Comparing the results of Eqs (9.6) and (9.7) it may occur that, depending on the loading history of the sample, the void ratio after collapse (Eq. (9.6)) can be smaller than the void ratio of a saturated sample loaded to the same mean net stress reached at the end of collapse (Eq. (9.7)). This means that close to the end of collapse, the soil sample may cross the saturated virgin consolidation line resulting in a slightly overconsolidated material. This behavior is generated by the drying-loading--wetting path followed by the collapsed sample (path ABEH) when compared with the simple loading path followed by the saturated sample (path AH). Moreover, when the collapsed sample crosses the saturated virgin consolidation line its behavior changes from elastoplastic to elastic as it invades the overconsolidated zone. This type of behavior has been experimentally observed by different authors (Sun, Matsuoka and Xu [126]; Pereira, Fredlund, Cardao Neto and Gitirana [127]; Sun, Sheng and Xu [102]) and can be reproduced by the model presented herein because the geometry of the LCYS takes account of the whole drying-loading-wetting path followed by the sample (see Eqs. (9.2) and (9.3)). It is also possible to compute the suction at which the transition from elastoplastic to elastic behavior occurs (s_{DYS}) by equalizing the void ratio during collapse with the void ratio of the saturated sample at the end of loading (Eq. 9.7) resulting in

$$s_{DYS} = \frac{AB\left[e_I^{(1/\lambda_e)} - e_A^{(1/\lambda_e)}\right] + \chi_{wc}s_c\,C}{\chi A(e_A^{(1/\lambda_e)} - e_I^{(1/\lambda_e)}) + \chi_w C}$$

where $A = P_{fab} + \bar{p}_0 + \Delta\bar{p}_0 + \chi_0 s_0$, $B = \chi_0 s_0 - \chi_{00}s_{00} + p_{fab}$ and $C = p_{fab} + 2\chi_0 s_0 - \chi_{00}s_{00}$. The value of s_{DYS} represents the position of the SDYS at the specific effective mean stress. It is noteworthy that the position of this surface depends on the whole history of loading and suction imposed to the sample.

9.4. NUMERICAL AND EXPERIMENTAL COMPARISONS

In this section, the proposed framework for the volumetric behavior of unsaturated soils has been used to simulate the response of samples compacted at different densities and subjected to loading-wetting paths.

Sun, Sheng, Cui and Sloan [120] reported a series of suction controlled triaxial tests to observe the influence of the initial density on the collapse of compacted soil samples of Pearl clay. The specimens were prepared by static compaction in five layers at a vertical stress of 0.3 MPa, 0.4 MPa or 0.6 MPa. This procedure resulted in void ratios ranging from 1.0 to 1.5 and suctions ranging from 0.09 to 0.13 MPa. The volumetric deformation of the sample was obtained by measuring its lateral and vertical displacements. The lateral displacements were measured at three different heights of the sample and the volume was obtained by approaching the lateral shape of the sample to a third order polynomial equation. After compaction, all samples were subject to an equalization stage by applying a suction of 0.147 MPa and an isotropic mean stress of 0.02 MPa. Then the specimens were isotropically loaded to a previously specified net stress under a constant suction of 0.147 MPa. Finally, suction was decreased by steps from 0.147 to 0 MPa maintaining the nets stress constant. During this wetting stage, the SWRCs for the different densities and for each isotropic stress were obtained. Finally, the grain size distribution (GSD) of the material was also reported.

In order to be able to simulate the volumetric behavior of soils within the framework proposed in this paper, the values of parameter χ need to be determined according to the wetting-drying history of the soil. To this purpose the porous-solid model is used along with the SWRC of the material. As explained before, by fitting the main wetting and drying SWRCs of the material it is possible to determine the pore size distribution (PSD) of cavities and bonds. Then it is possible to simulate any drying-wetting process and define the parameters f^s, f^u and S_w^u required to determine the value of parameter χ along any drying-wetting path. Unfortunately, during the implementation of these tests only the wetting SWRC of the material was obtained. Therefore, only the wetting branch could be fitted with the experimental points while the drying branch has been guessed.

Fig. (**4a**) shows the fitting of the wetting branch of the SWRC and the guess made for the drying branch for a sample compacted at a void ratio of e = 1.33. From this fitting, the PSD of cavities and bonds for this specific density of the material could be found. Fig. (**4b**) shows the cavity size distribution of samples compacted at different void ratios (including that for e=1.36) and their comparison with the GSD of the material. In this figure the reduction in cavity sizes for the different compacted void ratios can be observed.

Once the PSD of the material has been established, it is possible to determine the values of parameters f^s, f^u, f^d and S_w^u by simulating the drying and wetting process of the soil. These parameters are shown in Fig. (**4c** and **4d**) for the drying stage (reaching a maximum suction of 0.15 MPa) and the subsequent wetting stage, respectively. Using Eq. (2.15) (Chapter 2), it is now possible to define the

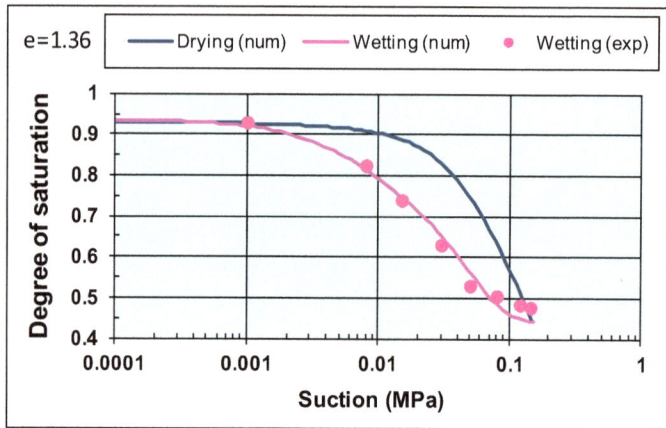

(a)

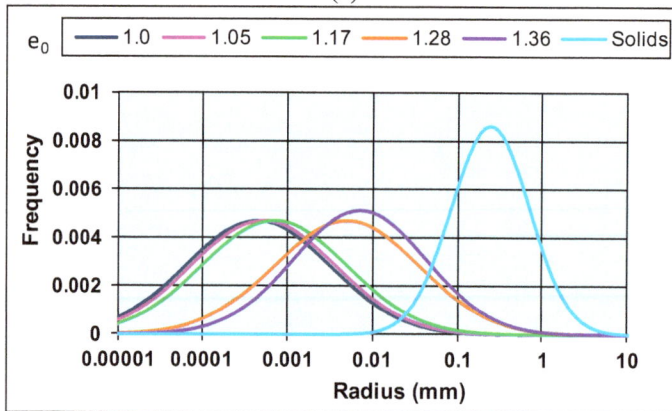

(b)

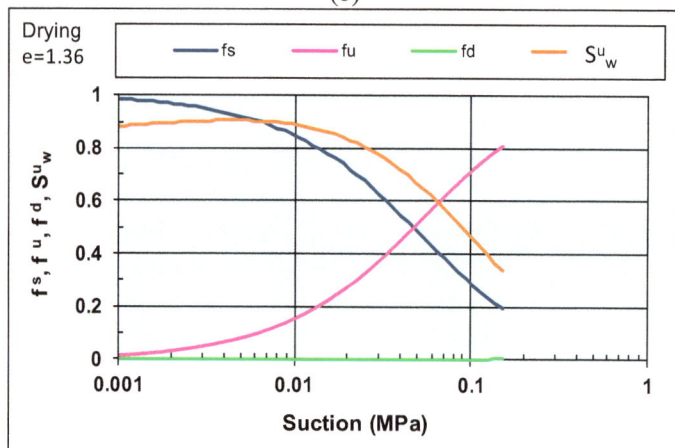

(c)

Fig. 1 cont.....

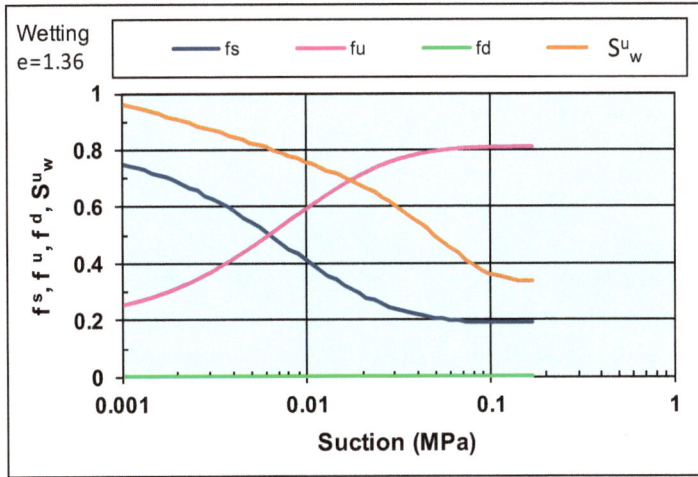

(d)

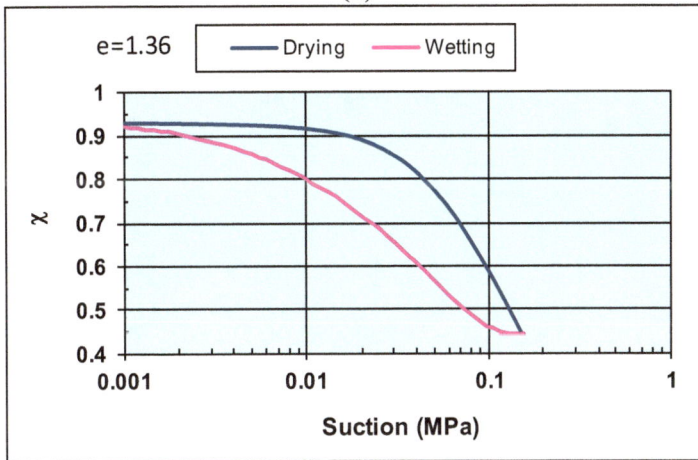

(e)

Fig. (4). Numerical simulations obtained with the porous-solid model. (**a**) Fit of the wetting curve, (experimental data from [122]) (**b**) cavity size distribution for different densities and GSD of the material, (**c**) and (**d**) parameters f^s, f^u, f^d and S_w^u at drying and wetting, respectively and (**e**) values of parameter χ at drying and wetting.

value of parameter χ during the drying and wetting stages. These values are shown in Fig. (**4e**) as function of suction. Once this parameter has been defined it is now possible to determine the effective stresses at any stage of the test (Eq. (2.14), Chapter 2), simulate the behavior of the material at loading (Eq. (8.4), Chapter 8), define the geometries of the LCYS and the WP (Eqs. 9.3 and 9.4, respectively), define the point at which these curves intersect and determine the volumetric response of the soil during collapse (Eqs. (8.4), (8.5), (9.4) and (9.5)). This same procedure has to be done for each density of the material and for each value of the mean net stress.

9.4.1. Isotropic Compression

The comparison between experimental and numerical results for samples compacted at different densities and subject to isotropic loading is presented in Fig. (**5**). The average initial void ratios of tested samples were: 1.36, 1.28, 1.17 and 1.05. Curves for both unsaturated and saturated conditions are presented. These curves were obtained using Eq. (8.4, Chapter 8) with $\lambda_e = -0.12$. Arrows in this figure indicate the value of the fabric stress for each density. The fabric stress for the samples compacted at void ratios of 1.36, 1.28, 1-17 and 1.05 were 0.03, 0.06, 0.09 and 0.15, respectively. In general, there is good agreement between experimental and numerical results although for samples compacted at a void ratio of 1.28, the results are not as good as for the other samples.

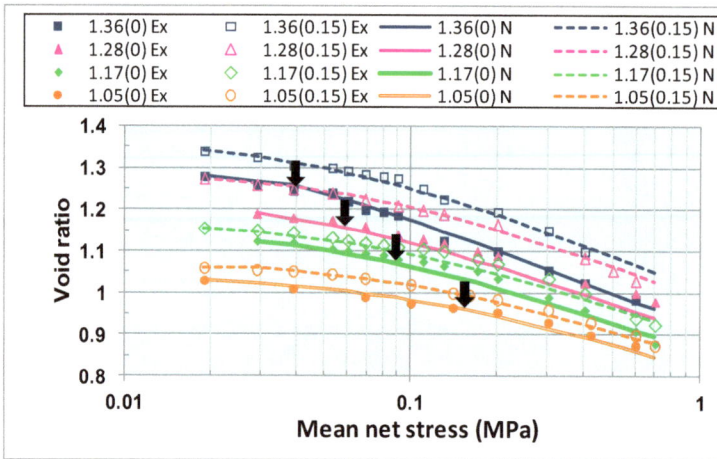

Fig. (5). Isotropic loading tests on samples compacted at different densities. The first number identifying the test indicates the "as compacted" void ratio and the number in parenthesis indicates the value of suction during compression in MPa. Ex and N stand for experimental and numerical results, respectively (experimental data from [122]).

9.4.2. Collapse Upon Wetting

Figs. (**6-9**) show some experimental and numerical comparisons for soils samples compacted at four different densities (e_0=1.05, 1.17, 1.28 and 1.36, respectively) then loaded to different isotropic stresses and finally subjected to a wetting path. In these figures, plots (a) represent the shapes of both the LCYS and the WP in the axis of effective stress *versus* suction for samples compacted at the same density and loaded at different isotropic stresses (ranging from 0.02 to 0.6 MPa). Plots (b) represent the volumetric response of the soil during the wetting stage in the axis of suction *versus* volumetric strain. Finally, plots (c) represent the collapse volumetric strain of the sample *versus* the applied mean net stress. The LCYS and the WP were obtained from Eqs. (9.3) and (9.4), respectively. In plots (a) and (b) an initial elastic rebound of the material can be observed followed by the collapse. This type of behavior has also been experimentally observed by Rodrigues and Volar [121]. The value of the unloading-reloading slope considered for these simulations was $\kappa_e = -0.01$. It can also be noticed that depending on the isotropic net stress applied to the sample, the WP crosses the LCYS at different stages of the wetting process. The larger the applied isotropic net stress during the loading stage, the earlier the phenomenon of collapse appears during the wetting stage. For the sample compacted at a void ratio of 1.05 and loaded to 0.02 MPa, collapse never occurs. Instead, for samples compacted at void ratios of 1.17, 1.28 and 1.36 and loaded at the same isotropic stress, collapse only occurs close to saturation. This happens because the net stress applied to these samples, $\bar{p}_0 = 0.02\ MPa$, is close to the value $\chi_0 s_0 - \chi_{00} s_{00}$ representing the hardening of the LCYS during the equalizing stage.

The start of collapse and the evolution of the volumetric deformation of the soil during collapse are correctly simulated by the model (see Figs. **6b**, **7b**, **8b** and **9b**). However, for samples compacted at large void ratios ($e = 1.28$, 1.36) and subjected to large isotropic stresses, the slope of the numerical volumetric strain against suction remains low compared with the experimental results. This causes under predicted volumetric strains for large values of the mean net stress. In spite of these differences, it can be observed that numerical simulations correctly predict a maximum collapse volumetric strain with increasing mean net stress (Figs. **6c**, **7c**, **8c** and **9c**).

As stated before this model includes the hysteresis of the SWRC and to some extent the hydro-mechanical coupling of unsaturated soils as the reduction on the size of cavities due to loading or suction increase has not been included.

The data required by the model to produce these simulations are: the main drying and wetting SWRCs, the GSD of the material, the fabric stress p_{fab}, the slope of both the virgin consolidation line (λ_e) and the unloading-reloading line (κ_e) in the axes of void ratio *versus* the logarithm of the mean net stress, the initial mean net stress ($\bar{p}_0$), suction (s_0) and voids ratio of the material (e_0) and the previous loading history or fabrication process in order to define the initial position of the LCYS.

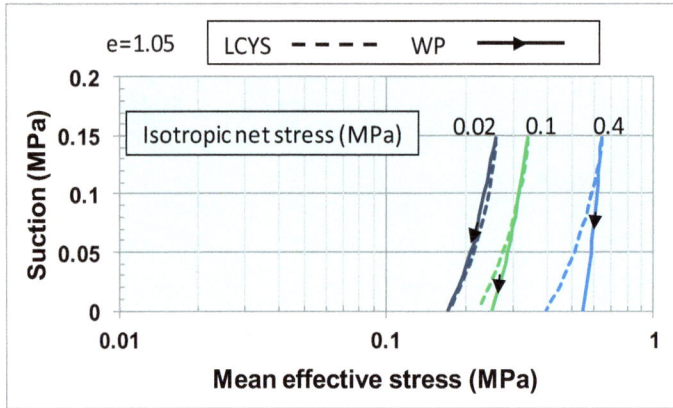

(a)

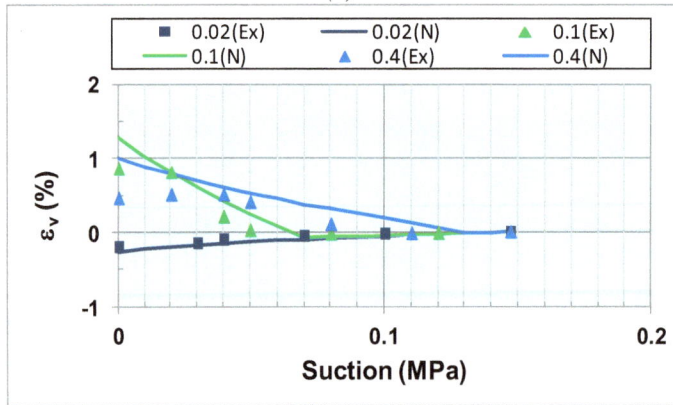

(b)

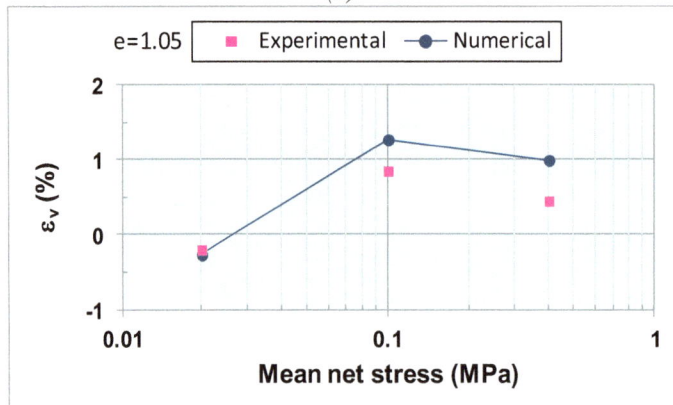

(c)

Fig. (6). Samples compacted at an initial void ratio e_0=1.05. (**a**) LCYSs and WPs, (**b**) volumetric strain against suction and (**c**) volumetric strain against mean net stress (experimental data from [122]).

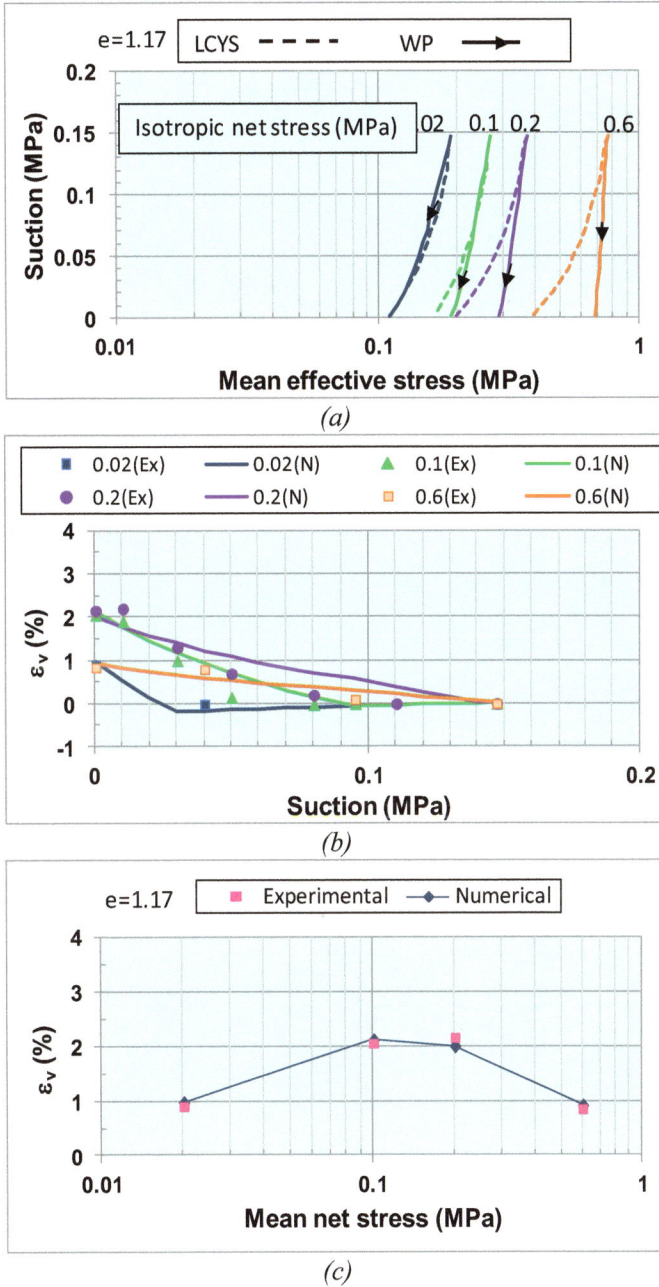

Fig. (7). Samples compacted at an initial void ratio e_0=1.17. (**a**) LCYSs and WPs, (**b**) volumetric strain against suction and (**c**) volumetric strain against mean net stress (experimental data from [122]).

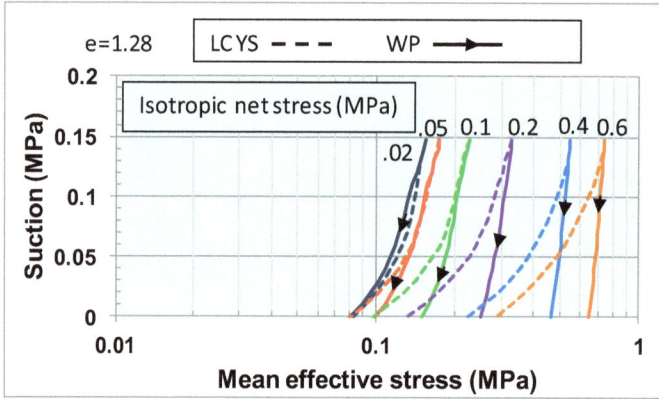

(a)

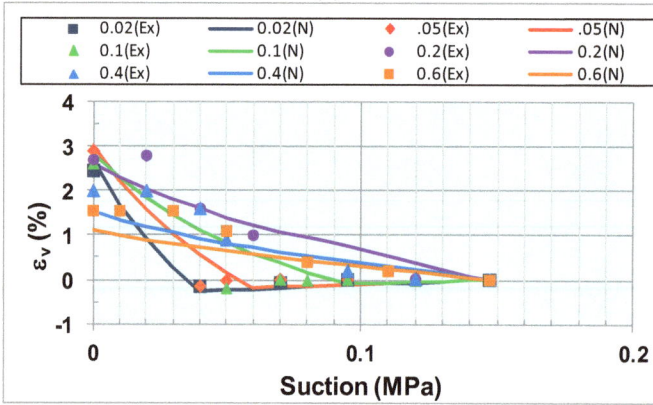

(b)

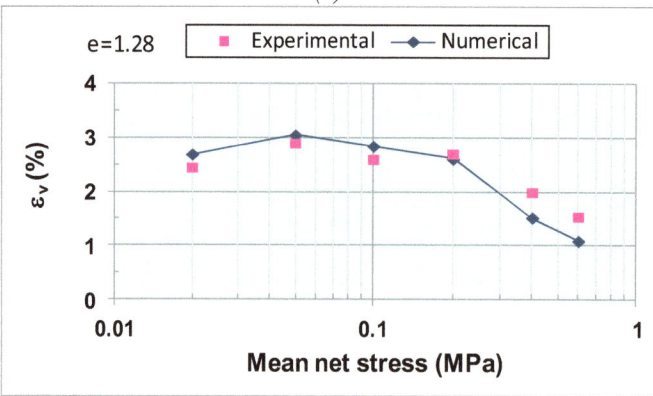

(c)

Fig. (8). Samples compacted at an initial void ratio $e_0=1.28$. (**a**) LCYSs and WPs, (**b**) volumetric strain against suction and (**c**) volumetric strain against mean net stress (experimental data from [122]).

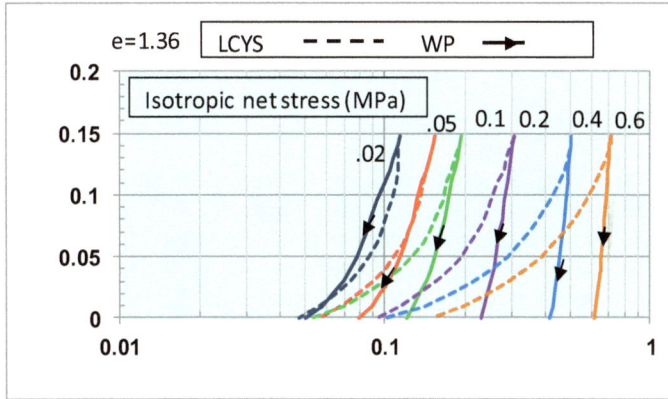

(a)

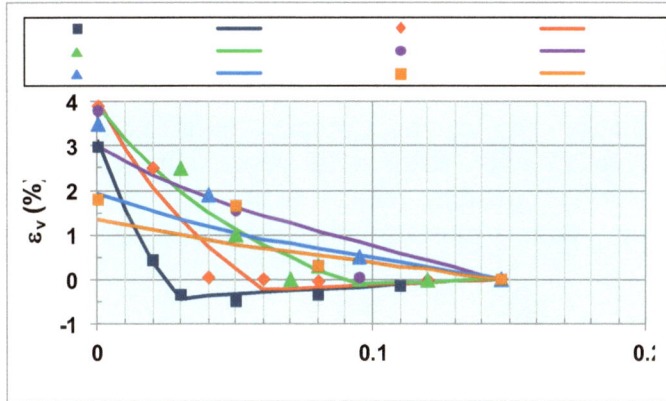

(b)

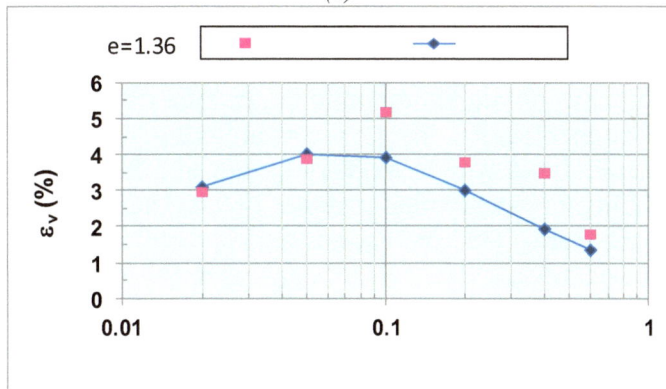

(c)

Fig. (9). Samples compacted at an initial void ratio e_0=1.36. (**a**) LCYSs and WPs, (**b**) volumetric strain against suction and (**c**) volumetric strain against mean net stress (experimental data from [122]).

CHAPTER 10

Expansive Soils

Abstract: In this chapter, an elastoplastic framework for the volumetric behavior of expansive soils based on effective stresses is proposed. This framework is grounded on the volumetric and collapsing behavior of soils developed in Chapters 8 and 9. The hydraulic behavior of the soil is simulated using the porous-solid model developed in Chapter 4. The result is an elastoplastic framework based on the equation of the volumetric behavior of saturated soils presented in Chapter 8 where the value and sign of the compression index depends on the position of the state of stresses and the direction of the increment of the effective stress with respect to the yield surfaces in the plane of effective mean stress against suction. Experimental and numerical comparisons show the ability of the model to simulate the behavior of expansive soils under different stress paths.

Keywords: Collapse upon loading, Compression index, Compression-expansion index, Drying path, Effective stresses, Elastoplasticity, Expansive soils, Loading collapse yield surface, Loading path, Relative density, Suction increase yield surface, Unloading path, Unloading-reloading index, Wetting path.

10.1. INTRODUCTION

It is known that montmorillonite is one of the most active clays in nature. It appears as equidimensional flakes with a maximum length of 1 or 2 μm and a medium thickness of 0.001 μm for a single platelet or basic layer. Each basic layer is formed by one octahedral sheet packed between two silica sheets. The octahedral sheet may contain aluminum, magnesium, iron, zinc, nickel, lithium or other cations (Mitchell [128]). These sheets are bonded together in a basic layer by primary valence links which are very strong and hard to destroy. A certain number of basic layers can be piled together by electrical forces forming a single clay particle. In general, positive cations and molecules of water can make the link between basic layers. Their secondary specific surface (*i.e.* the one that considers the interlayer surface) may be as large as 800 m²/g. This means that electrical forces are of great importance for the behavior of these materials. Due to the bipolar characteristics of the molecule of water and depending on its availability, additional layers of water can intrude the link between the different platelets of a particle. When this happens, particles increase their volume and clay exhibits expansion. This phenomenon is called crystalline swelling. If salt is diluted in the invading water a double layer develops between particles and repulsive forces appear. This phenomenon is called osmotic or double layer swelling (Anandarajah and Amarasinghe [129]). In this paper, only the case of crystalline swelling is considered.

It is also known that many natural fine materials and dry of optimum compacted soils show a bimodal structure formed by a microstructure and a macrostructure. The microstructure is represented by the arrangement of small particles forming aggregates with pores of small size. These pores are called intra-aggregated pores. The relative arrangement of aggregates constitutes the macrostructure which exhibits large pores. These pores are called inter-aggregated pores or macropores. The microstructure and the macrostructure become apparent through the pore size distribution (PSD) of the soil which can be obtained from mercury intrusion porosimetry (MIP) tests as shown by Simms and Yanful [45]. These researchers also observed that the relative volume of macropores reduces while that of intra-aggregated pores increases in relation to the total volume of pores, when a soil sample is loaded or dried (Simms and Yanful [45, 46]).

This chapter is organized as follows: first, some of the most successful models for expansive soils are revised. Subsequently, the elastoplastic framework developed in Chapters 8 and 9 is extended for the case of expansive soils. Latter, some comparisons between numerical and experimental results are presented.

10.2. BACKGROUND

One of the most successful models to simulate the behavior of expansive soils was proposed by Gens and Alonso [130] and later improved by Alonso, Vaunat and Gens [131]. These authors developed an elastoplastic framework for the behavior of expansive soils based on the independent stress variables approach. This model includes the behavior of the two structural levels: the microstructure and the macrostructure. The microstructure is represented by swelling aggregates. This microstructure remains mostly saturated and its behavior is governed by the effective stress obtained from the addition of net stress and pore water pressure. In that sense, a change in suction or net stress of the same quantity results in the same volumetric strain of the microstructure. The macrostructure is represented by the particular arrangement of aggregates and coarse material. These elements form the inter-aggregated pores where major structural rearrangements occur. The macrostructure is highly affected by external loading or suction increase. However, the volumetric response of the macrostructure has low direct influence on the microstructure. In contrast, the macrostructure can be greatly affected by the volumetric behavior of the microstructure. When an expansive soil is subjected to wetting–drying cycles, the volume of its microstructure alternatively increases and reduces and, this volumetric behavior is partially transmitted to the macrostructure. Therefore, it can be said that there is a single direction coupling between microstructure and macrostructure. Gens and Alonso [130] consider that this coupling between micro and macrostructure depends on the overconsolidation ratio of the sample. Also these authors consider microstructural deformations as largely reversible while irreversible behavior is attributed to the macrostructure.

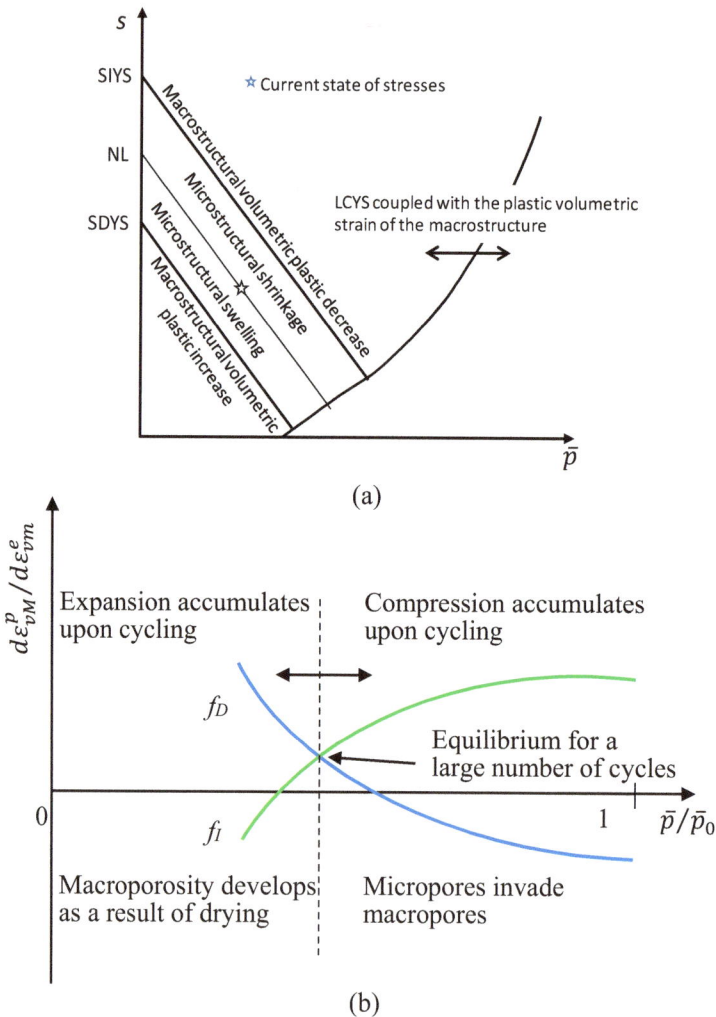

(a)

(b)

Fig. (1). (**a**) Yield surfaces for the Barcelona Expansive Model and (**b**) micro-macro-pore interaction. Adapted from [131].

The modeling of the macrostructure is described in the original Barcelona Basic Model (BBM) by Alonso, Gens and Josa [7]. The elastic region of the macrostructure in the plane of mean net stress ($\bar{p}$) *versus* suction (s) is bounded by the loading collapse yield surface (LCYS) as indicated in Fig. (**1a**). In order to include the behavior of expansive soils in this model, an expansive yield surface is added. This surface is called the Neutral Line (NL) and is represented by a straight line forming 135° with the horizontal in the plane formed by net stress and suction (see Fig. **1a**). When this yield surface is crossed by a suction or stress reduction path, the microstructure expands. A related movement between the NL and the LCYS takes into account the coupling between microstructure and macrostructure during expansion.

In order to include the macrostructural elastoplastic volumetric strains generated by the reversible microstructural strains, two additional yield surfaces were introduced: one is the suction increase yield surface (SIYS) and the other the suction decrease yield surface (SDYS) as indicated in Fig. (**1a**). These surfaces run parallel to the NL; the former is located above while the last is located below the NL. The SIYS and the SDYS bound the macrostructural elastic volumetric strains during shrinkage and swelling of the microstructure, respectively. In addition, the dependency of the volumetric elastoplastic strains of the macrostructure on the elastic strains of the microstructure was included by means of two functions: f_D and f_I for suction decrease or increase, respectively. These functions depend on the ratio $\bar{p}/\bar{p}_0$ (where $\bar{p}_0$ represents the apparent preconsolidation net stress) for isotropic stresses and their shape is plotted in Fig. (**1b**). This ratio has the same meaning as the overconsolidation ratio for saturated soils. A possible choice for these functions is given by Alonso, Vaunat and Gens [131]:

$$f_I = f_{I0} + f_{I1}\left(\tfrac{\bar{p}}{\bar{p}_0}\right)^{n_I} \text{ and } f_D = f_{D0} + f_{D1}\left(1 - \tfrac{\bar{p}}{\bar{p}_0}\right)^{n_D}$$

where $f_{I0}, f_{I1}, n_I, f_{D0}, f_{D1}$ and n_D are parameters of the model. Another possibility is given by Lloret, Villar, Sánchez, Gens, Pintado and Alonso [132]

$$f_I = f_{I0} + f_{I1} \tanh\left|n_1\left(\tfrac{\bar{p}}{\bar{p}_0}\right) - n_2\right| \text{ and } f_D = f_{D0} - f_{D1} \tanh\left|n_1\left(\tfrac{\bar{p}}{\bar{p}_0}\right) - n_2\right|$$

where $f_{I0}, f_{I1}, n_1, n_2, f_{D0}$ and f_{D1} are parameters of the model. When the SDYS is activated, function f_D is used and the increment of the macrostructural volumetric strains is obtained from the equation

$$d\varepsilon_{vM}^p = f_D \, d\varepsilon_{vm}^e$$

On the contrary, when the SIYS is activated, function f_I replaces function f_D in the above equation. The volumetric strains transmitted from the microstructure to the macrostructure affect the void ratio of the sample and therefore the position of the LCYS. However, volumetric plastic strains generated by loading modify the position of the LCYS only, not the position of the SIYS or SDYS.

In a latter version of the model, hydromechanical coupling between the micro and macrostructure was included in order to study the transient flow of water (Alonso, Romero and Hoffmann [133]). In such a case, different void ratios, retention curves, saturated permeability, degrees of saturation and suctions are used for the micro and macrostructure of the soil. This model has been successful in simulating the behavior of expansive soils following a variety of stress paths and even it can simulate the evolution of suction and volumetric strains with time during saturation.

Using a different approach, Sun, Sun and Li [134] and Sun and Sun [135] proposed a constitutive model to simulate the behavior of expansive soils based on the effective stress concept. This model extends an elastoplastic framework for unsaturated soils and includes fully hydro-mechanical coupling. It adopts the Bishop's effective stress equation with Bishop's parameter χ equal to the degree of saturation ($\chi = S_w$). The retention curves are simulated using an elastoplastic model similar to that proposed by Wheeler, Sharma and Buisson [22] except that in this case, the air entry value is dependent on the void ratio to include hydro-mechanical coupling. In this framework, the apparent preconsolidation stress increases with suction in a similar way to the BBM. It establishes a reducing compression index as suction increases. It also considers a linear relationship between the degree of saturation and the void ratio when compression tests at constant suction are performed; also, this linear relationship shows the same gradient for different suctions. This model seems to have great potential to simulate the behavior of expansive soils although, in the author's knowledge, it has not been confronted with experimental loading-unloading or wetting-drying tests.

Other interesting models based on effective stresses have been proposed by Mâsín [136], Della Vecchia, Jommi and Romero [121] and Li, Yin and Hicher [137]. These models are similar to the one proposed by Alonso, Romero and Hoffmann [133], in the sense that they consider two independent coupled hydromechanical models, one for the macrostructure and the other for the microstructure. The one by Mâsín [136] even includes independent measures of effective stresses in each structure. The global response of the soil is obtained by a coupling function that links both structural levels. The model proposed by Li, Yin and Hicher [137] uses work conjugate stress-strain variables. These models correctly simulate the behavior of expansive soils during wetting-drying cycles although this two structural levels approach requires a large number of parameters.

This chapter presents an elastoplastic framework for expansive soils based on effective stresses and the equation of volumetric behavior of saturated materials. This framework does not require any additional yield surface to include expansion and uses only two additional compression indexes in addition to the loading and unloading-reloading saturated compression indexes. The first depends on the last and the relative density of the material. It uses global parameters and therefore, does not require splitting the soil in two structural levels and includes the hydromechanical coupling of unsaturated soils. The result is a simpler model with less parameters and similar precision to other models. The behavior of expansive soils compacted at different densities during loading-unloading and wetting-drying cycles can be fairly simulated. It is also used to explain the behavior of soils during constant volume expansive pressure tests. Note that this chapter only develops the elastoplastic framework for the volumetric response of the soil. In the next chapters, this framework is incorporated in a fully coupled elastoplastic constitutive model for unsaturated soils to include the influence of the deviator stress.

10.3. EXTENDED ELASTOPLASTIC FRAMEWORK

The elastoplastic framework for expansive soils developed by Alonso, Vaunat and Gens [131] has been quite successful in modeling the behavior of these materials and is adopted herein as the basis to develop an elastoplastic framework based on effective stresses. Therefore, it is considered here that the microstructure swells during wetting or unloading and shrinks during drying or loading and this behavior is transmitted to the macrostructure somehow.

According to the experimental results by Blight [138] and Kassiff and Shalom [139], when an expansive soil is wetted under constant volume conditions, the slope of the stress path at the saturated condition ($\chi = 1$) in the axes of suction *versus* swelling pressure (net stress) shows a value of 45°. This means that the swelling index takes the same value as the compression index. This consideration is also similar to the concept of "equivalent matric suction" proposed by Fredlund, Hasam and Filson [140] and Vu and Fredlund [141] and adopted by Tu and Vanapalli [142]. At the present stage of the model, the effect of adsorbed water layers is disregarded.

When the aggregates swell upon wetting, part of this swelling is transmitted to the macrostructure. Then, as stated by Gens and Alonso [130], an interrelationship between the microstructure and the macrostructure has to be included in the model. This interrelationship needs to account for the fact that when water is available and the microstructure swells, this increase in volume is not fully transmitted to the macrostructure as some of this expansion occurs inside the macropores (Monroy, Zdravkovic and Ridley [143]). Therefore, the initial volume of micro and macropores (which defines the density of the soil) becomes an important parameter for expansion modeling. This can be observed in Fig. (2) where the experimental (Ex) saturated vertical strains for different dry densities and vertical loads have been plotted as a function of the void ratio (Hoffmann, Alonso, Romero [144]; Villar and Lloret [145]). The graph results in linear relationships between the expansive strain and voids ratio for a constant vertical stress. If these results are extrapolated, a void ratio for nil expansive strain can be defined. This value is considered here as the maximum possible void ratio (e_{max}) that a sample can reach. This value can be obtained at the end of a free expansion test when the soil is fully saturated.

It is also considered here that there is a void ratio that ensures the full transmission of the expansive strains from the microstructure to the macrostructure. In that case, the macropores practically disappear from the soil structure (Monroy, Zdravkovic and Ridley [143]) and then, all the soil mass shows the density of aggregates. This value represents the minimum possible voids ratio (e_{min}) of the soil reached during heavy compaction at the optimum water content. This procedure reduces the macrostructure to a minimum. Under these considerations, the numerical (N) results shown in Fig. (2) have been obtained.

Accordingly, it can be stated that the plastic volumetric strain transmitted by the microstructure to the macrostructure is a function of the relative density of the sample ($D_r = (e_{max} - e)/(e_{max} - e_{min})$) while the elastic behavior is not affected by density. A similar parameter has been considered by Mâsín [136] to evaluate the influence of a swelling microstructure on the macrostructure. Then, to account for the elastoplastic volumetric expansive behavior of the macrostructure, the following expansion index is used

$$\lambda_{ex} = (\lambda_e - \kappa_e)D_r + \kappa_e \tag{10.1}$$

In addition, when an expansive soil subjected to a certain suction and sufficiently loaded is then saturated, macrostructural collapse occurs while microstructural expansion takes place at the same time (Alonso, Romero and Hoffmann [133]). In such a case, the modeling of the soil needs to take account of both phenomena simultaneously. This is done by considering a collapse-expansion index (λ_{cex}) resulting from the difference between the compression index and the expansion index (Eq. 10.1) added by the elastic index, in the form

$$\lambda_{cex} = \lambda_e - \lambda_{ex} + \kappa_e = (\lambda_e - \kappa_e)(1 - D_r) + \kappa_e \qquad (10.2)$$

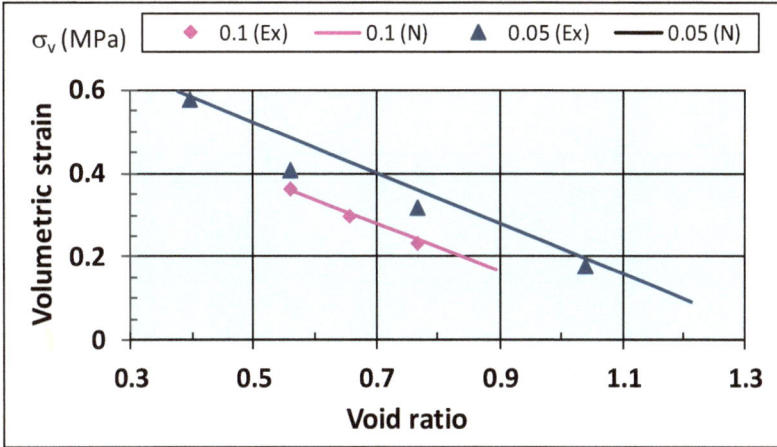

Fig. (2). Expansive strains for samples with different densities loaded at 0.1 and 0.05 MPa. Experimental results from [145] and [144], respectively.

When the wetting path (WP) crosses the LCYS, the resulting volumetric strain is compressive unless $D_r = 1$ (for most cases $\lambda_e > \lambda_{ex}$).

Another interesting phenomenon occurs when measuring the expansion pressure of a sample during a constant volume wetting test as that performed by Romero, Gens and Lloret [109] and shown in Fig. (3). Initially, the sample tends to increase its volume as suction reduces and at the same time, the mean net stress increases to maintain the volume constant (stage 1, Fig. 3). However, at certain moment the WP meets the LCYS. As this surface represents the natural path for null volumetric strain (neutral loading surface) then, the stress path follows the LCYS and the mean net stress adjusts to the required value (stage 2). If a drying-wetting cycle is applied at the end of saturation, the stress path simply comes back along the LCYS when plotted in the effective stress plane as will be shown latter (stage 3). Although different interpretations have been given to these results (Gens and Alonso [130], Romero, Gens And Lloret [109], Lloret, Villar, Sánchez, Gens, Pintado and Alonso [132]), this behavior can be simulated under the effective stress framework as explained below.

The first stage of the test is simulated by computing the increment of the mean net stress required to maintain a constant volume in a sample that wets from an initial state of stresses. To this purpose two steps are followed. First, it is considered that the soil expands due to a small decrement of suction and second, the increment of the net stress required to recompress the soil to its original volume is calculated. During the first step it is considered that the net stress remains constant while suction changes and the soil expands. The volumetric strain of the sample produced by wetting is computed using Eq. (8.3) (Chapter 8), in the form

$$\Delta\varepsilon_v = \frac{e_0\left[\left(\dfrac{p_0' + \Delta(\chi s)}{p_0'}\right)^{\lambda_{ex}} - 1\right]}{1 + e_0} \tag{10.3}$$

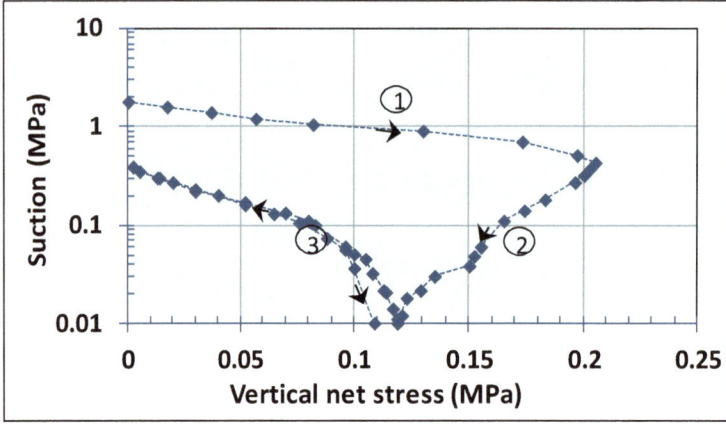

Fig. (3). Constant volume swelling pressure test with wetting-drying cycles in a sample with dry density of 13.7 kN/m³. Adapted from [109].

In this case, $\Delta(\chi s)$ represents the variation in the effective stress produced by wetting. Note that the expansion index λ_{ex} is used in the above relationship. In the second step, it is considered that suction remains constant while the mean net stress increases to recompress the soil. The increment of the mean net stress required to recompress the sample to its original volume is computed also with Eq. (8.3) (Chapter 8), in the form

$$\Delta\bar{p} = \frac{p'(1 + e_0)}{e_0\,(\lambda_{cex})}\Delta\varepsilon_v \tag{10.4}$$

Observe that the compression index λ_{cex} is used in the above equation as it is considered that the soil expands first and then recompress. By substituting Eq. (10.3) in (10.4) it is possible to compute the mean net stress required to perform a constant volume wetting test until the stress path meets the LCYS.

During the second stage, the stress path reaches and follows the LCYS. The LCYS represents the path for null volumetric strain change and therefore for null effective stress change $dp' = d\bar{p} + d(\chi s) = 0$. Accordingly, the variation in net stress equals the variation in suction stress

$$d\bar{p} = d(\chi s) \tag{10.5}$$

During loading-unloading tests, the soil behaves according to the general elastoplastic framework for unsaturated soils presented in the previous section. For every plastic volumetric strain generated during wetting-drying cycling, the LCYS hardens or softens depending on the sign of the volumetric strain. This hardening or softening can be computed using Eq. (8.3) (Chapter 8) in the form

$$\Delta p' = p' \left(\frac{\Delta e^p}{e}\right)^{\left(\frac{1}{\lambda_e - \kappa_e}\right)} \tag{10.6}$$

Additionally, during wetting-drying cycles and just after the inversion, an initial elastic behavior is observed. It is therefore important to include a SDYS in addition to the SIYS, in the same way as proposed by Alonso, Vaunat and Gens [131]. The SDYS hardens in such a way that the elastic behavior becomes predominant after few regular wetting-drying cycles (Nowamooz and Masrouri [146]). This last phenomenon can be explained by the fact that with further cycles, clay particles may rotate or displace in such a way that the swelling of the microstructure is accommodated inside the pores of the macrostructure.

In brief, the elastoplastic framework for expansive soils proposed herein has the following features:

a. The macro-structural behavior is based on the elastoplastic framework developed to simulate the volumetric behavior of unsaturated soils.
b. During wetting-drying tests, the micro-structural behavior is regulated by matric suction and is represented by the expansion and shrinkage of clay platelets. The expansion index is related to the saturated virgin compression index and the relative density of the soil as defined by Eq. (10.1).
c. When the WP crosses the LCYS, the macrostructure experiences collapse while the microstructure undergoes expansion. In such case, the global elastoplastic compression index (λ_{cex}) is determined from the addition of the elastic index and the difference between the compression (λ_e) and the expansion (λ_{ex}) indexes as expressed in Eq. (10.2).
d. A SDYS is introduced in the elastoplastic volumetric framework in addition to the SIYS. In this way, the elastic zone for wetting-drying cycles is bounded and the initial elastic behavior of the macrostructure after a suction inversion can be simulated. This elastic zone grows with the number of wetting-drying cycles. However, as a first approximation for the model, it is considered here that the volumetric behavior of the macrostructure becomes elastic after the first regular wetting-drying cycle. In view of the evaluation of the present

model, more realistic assumptions can be incorporated latter to increase the precision during wetting-drying cycles.

e. The preconsolidation stress of the sample modifies as the soil shows plastic expansion or compression according to Eq. (10.6), *i.e.* the LCYS is coupled to the SIYS and SDYS.

This elastoplastic framework is represented in Fig. (**4**) in the plane of mean effective stress against suction. The elastic zone for wetting-drying cycles is represented by the area bounded by the SIYS and the SDYS. The elastic zone for the macrostructure is bounded by the LCYS and the SIYS.

f. The elastoplastic framework includes full hydro-mechanical coupling as the PSD continuously modifies with the plastic volumetric strain of the soil. This coupling does not require additional parameters or calibration process.

g. The parameters required by the model are: the loading and unloading compression indexes λ_e and κ_e, the retention curves in wetting and drying, the current as well as the maximum and minimum possible void ratio of the soil, the maximum suction applied to the sample and the preconsolidation stress in saturated conditions.

This development contributes to the construction of a unified soil mechanics theory for saturated and unsaturated soils.

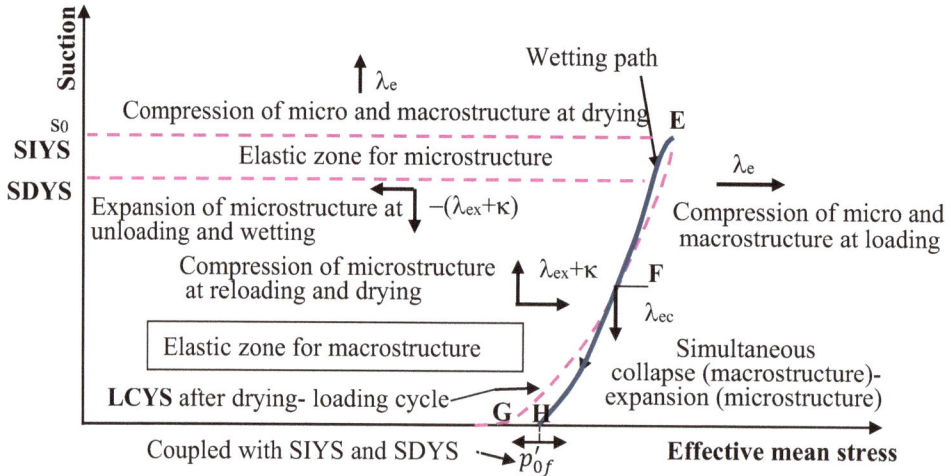

Fig. (4). Elastoplastic framework for expansive soils based on effective stresses.

10.4. EXPERIMENTAL AND NUMERICAL COMPARISONS

Romero, Gens and Lloret [109] performed a series of tests on statically compacted samples of Boom clay from Mol (Belgium). This clay shows a liquid limit $w_L = 56\%$, a plastic limit $w_p = 29\%$ and a specific weight of solids of 2.65. Samples statically compacted at a water content of 15% were prepared at two dry volumetric weights: $\gamma_d=13.7$ and 16.7 kN/m³. The as compacted void ratio, degree of saturation, suction and compaction stress for these samples are shown in Table **1** for both densities. Fig. (**5**) shows the experimental soil-water retention curves for samples compacted at both densities and measured during constant volume swelling tests. It also shows the numerical best fit from the porous-solid model. With this fitting, the numerical PSD shown in Fig. (**6**) is obtained for both densities. This figure also shows the PSD obtained from MIP tests performed on the samples. The experimental PSD of low density samples exhibits three peaks located at radius sizes of 0.03, 1.0 and 6.0 μm while the high density samples only exhibits two: one located at 0.007 and the other at 0.4 μm. It can be observed that the numerical PSD approximately reproduces the PSD obtained during MIP tests although, in general, the numerical results show smaller pore sizes.

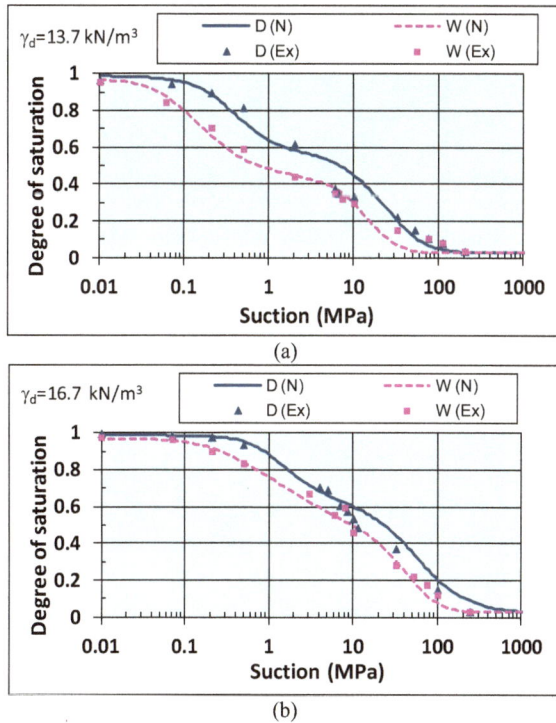

(a)

(b)

Fig. (5). Numerical fitting of experimental SWRCs at wetting and drying for (**a**) low density and (**b**) high density samples. Experimental data from [66].

Table 1. As compacted characteristics of Boom clay (adapted from [109]).

Density (kN/m³)	Void Ratio	S_w	Suction (MPa)	Compaction Mean Stress (MPa)
1.37	0.932	0.435	1.9	0.73
1.67	0.591	0.687	1.9	2.67

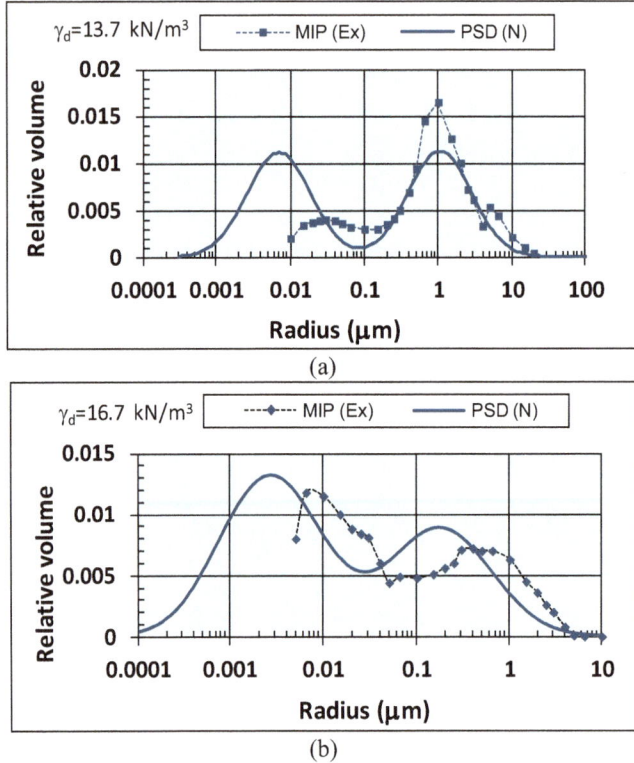

(a)

(b)

Fig. (6). Comparison of numerical and experimental pore size distributions for (**a**) low density and (**b**) high density samples. Experimental data from [66].

Fig. (**7a**) shows the values of parameters f^s, f^u, S_w^u and S_w for the loose sample ($\gamma_d = 13.7$ kN/m³) obtained from the porous-solid model at drying. These same parameters can be obtained during wetting-drying cycles and include the effect of hydromechanical coupling which modifies the retention curve depending on the evolution of the PSD of the sample. Observe that parameter χ shows the same limits and closely follows the evolution of the global degree of saturation S_w. This behavior of parameter χ ensures a smooth transition between saturated and unsaturated conditions at the same time that limits the effect of large suctions on the strength and volumetric behavior of unsaturated soils.

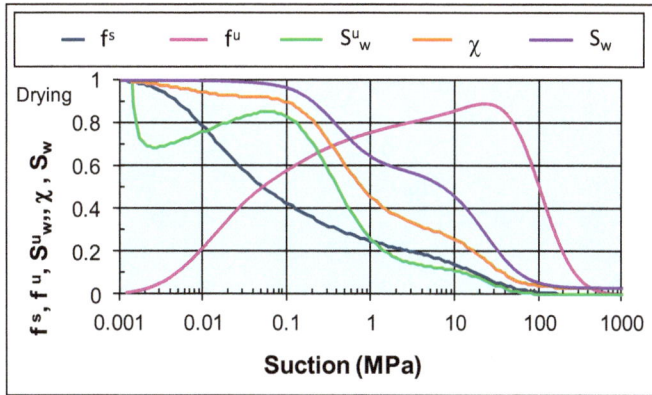

(a)

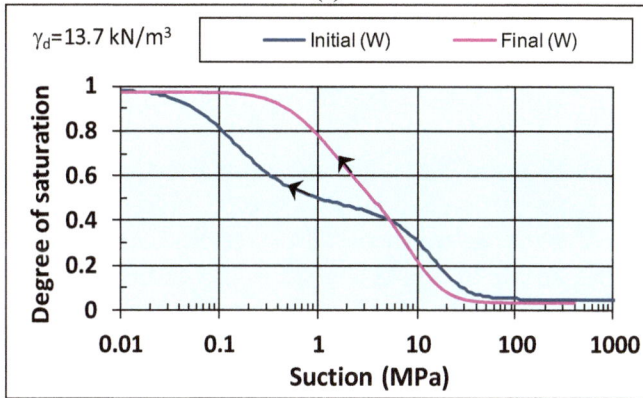

(b)

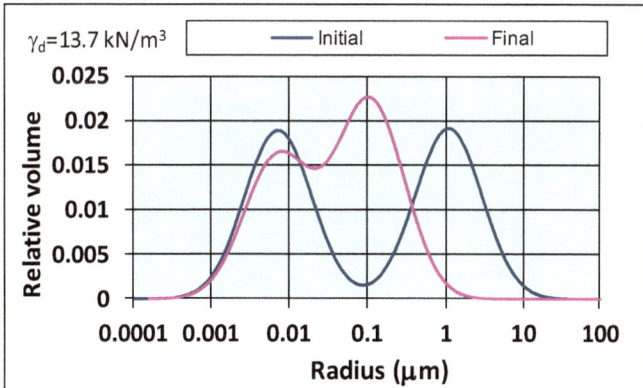

(c)

Fig. (7). (a) Values of parameters f^s, f^u, f^d, S_w^u and χ for the low density sample, **(b)** shift of the wetting retention curve and **(c)** change on pore size distribution after soil swelling.

According to MIP tests performed on swelling compacted samples of London clay subjected to wetting (Monroy, Zdravkovic and Ridley [143]), the inter-aggregated pores reduce both their frequency and size as the soil wets. In contrast, intra-aggregated pores increase their size and frequency. At full saturation, the inter-aggregated pores almost disappear while intra-aggregated pores increase their size in approximately one order of magnitude and four times their frequency. This same behavior has been observed by Romero, Della Vecchia and Jommi [147] on compacted samples of Boom clay. This means that macropores are being invaded by the swelling of mesopores and micropores. In that sense, the porous-solid model considers that the frequency and mean size of inter-aggregated pores gradually reduce as the soil wets while these same parameters increase for intra-aggregated pores. The frequency and size of these pores is determined from the current volume of the sample using a similar procedure as that established by Arroyo, Rojas, Pérez-Rea, Horta and Arroyo [148]. Using the updated pore size distribution, the porous-solid model defines the current retention curves of the swelling soil as it wets.

Fig. (7b) shows the shift of the wetting retention curve for the low density sample after the first wetting at a vertical stress of $\sigma_v = 0.02$ MPa. At this stage the sample shows a volumetric expansive strain of 11% (see Fig. 13a). Fig. (7c) shows the change in PSD of the sample at this stage according to the porous-solid model. In this example, the change in the PSD occurs by continuous expansion of the material with no collapse in the macrostructure (see Fig. 13b). In other cases shown later, the microstructure expands while the macrostructure collapses.

The tests performed by Romero, Gens and Lloret [109] included a series of controlled-suction tests in an oedometer cell applying wetting-drying cycles at different vertical net stresses in isothermal conditions. The samples were first loaded at constant water content from the as compacted state to a target vertical net stress. Then they were subjected to wetting-drying cycles with the following suction steps: 0.45, 0.2, 0.06 and 0.01 MPa. The results of these tests are shown in Fig. (8) for the low density samples and Fig. (9) for the high density samples. The vertical stresses applied to the former were: 0.085, 0.3 and 1.2 MPa while for the last: 0.026, 0.085 and 0.55 MPa. Also in these figures, the numerical simulations for each test are plotted. For the simulation of oedometric tests, a constant lateral earth pressure coefficient during soil expansion has been considered. This assumption is based on the experimental results reported by Habib [149]. This coefficient has been determined from loading tests reported by the authors and shown in Fig. (10).

In Fig. (8a) it can be observed that samples loaded at 0.085 and 0.3 MPa show an initial expansion (volumetric expansion is considered negative in this book) followed by collapse. This behavior is explained in Fig. (**8b**) where the initial LCYS and WP for each sample have been plotted. Observe that, for samples loaded at 0.085 and 0.3 MPa, the WP crosses the LCYS at suctions of 0.1 and 0.4 MPa, respectively. Therefore, collapse in these samples starts when suction reaches these values. Instead, for sample loaded at 1.2 MPa, the crossing of the LCYS occurs since the start of wetting and therefore this sample collapses from the beginning. Observe also that an important initial compressive strain occurs in this sample during the initial loading stage prior to wetting.

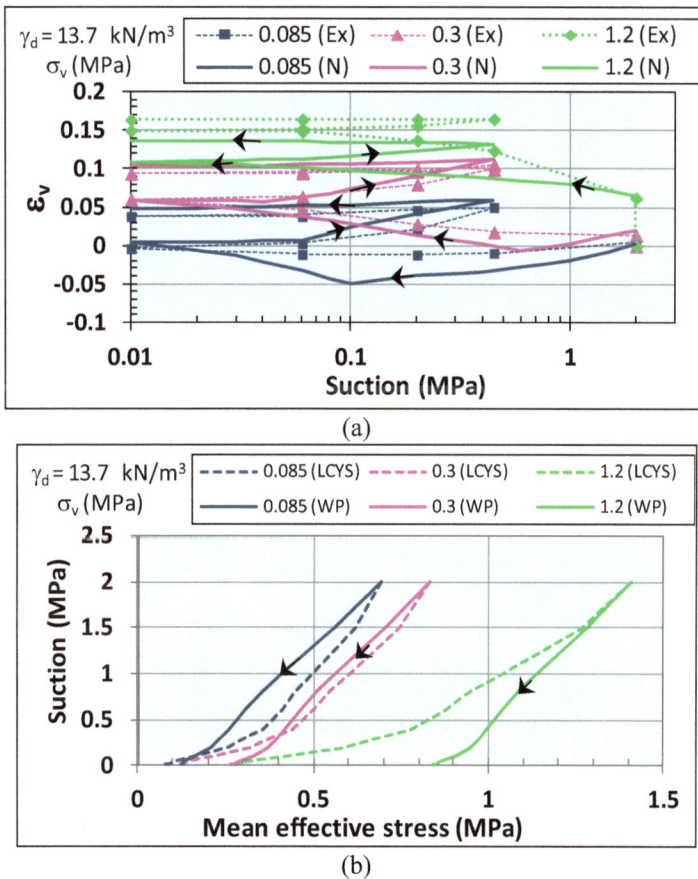

(a)

(b)

Fig. (8). (**a**) Volumetric behavior of loose samples during wetting-drying cycles and (**b**) initial LCYSs and WPs. Experimental data from [109].

With respect to samples compacted at high density (Fig. **9a**), it is observed that those loaded at 0.026 and 0.085 MPa have a continuous expansion while the one loaded at 0.55 MPa shows an initial expansion followed by collapse. This behavior is explained in Fig. (**9b**) where the initial LCYS and WP for each sample have been plotted. Observe that for samples loaded at 0.026 and 0.085 MPa, the WP never crosses the LCYS. Instead, for the sample loaded at 0.55 MPa, the WP crosses the LCYS at a suction of about 0.08 MPa, triggering the collapse of the sample.

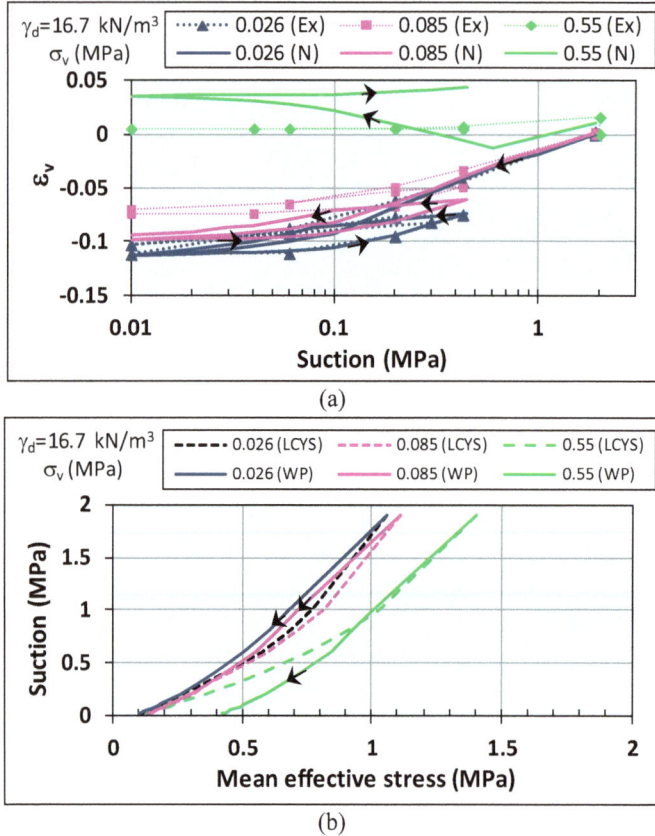

Fig. (9). (a) Volumetric behavior of dense samples during wetting-drying cycles and **(b)** initial LCYSs and WPs. Experimental data from [109].

Finally, during the first drying cycle after wetting there is an initial elastic behavior as the SDYS has been pulled downwards by the wetting path. When this elastic behavior ends, samples collapse again. This happens because when collapse has occurred during the first wetting, hardening takes place and the LCYS superposes now to the wetting path. However, the new stress path at drying appears on the right hand side of the LCYS because of hysteresis, producing additional collapse in the sample.

The numerical results correlate well with the experimental points during wetting-drying cycles for both densities except for the low density sample tested at a vertical stress of 1.2 MPa in Fig. (**8a**) and the high density sample loaded at 0.55MPa in Fig. (**9a**).

At the end of the wetting-drying cycles, samples were subjected to loading-unloading cycles. The volumetric behavior for both loose and dense samples is shown in Fig. (**10**) in the axes of mean effective stress against void ratio. From the simulation of these results the following values for the loading compression index for the low and high density samples were adopted: λ_e = -0.21 and λ_e = -0.16, respectively, while for the recompression index a value of κ_e = -0.025 for both densities was used. A lateral earth pressure coefficient of 0.39 resulted in the best fit for the experimental results and the same value was employed to simulate the expansion tests. Observe that results tend to align in a single compression curve for samples with the same density.

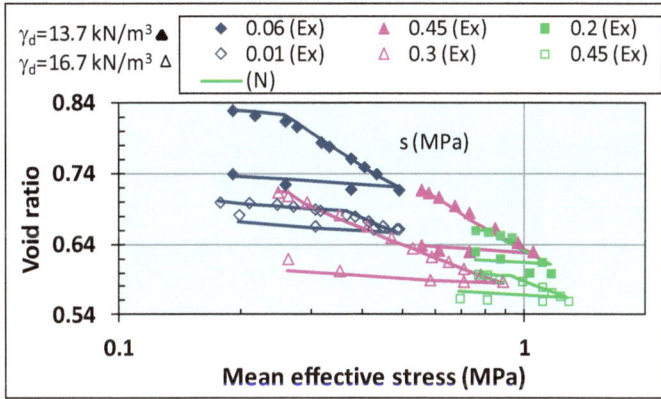

Fig. (10). Loading-unloading cycles for samples compacted at two densities after wetting-drying cycles. Experimental data from [109].

The relative density of samples was computed with the following maximum and minimum void ratios for both densities: e_{max} = 1.2 and e_{min} = 0.3. The maximum void ratio was established by extrapolating the experimental results (σ_v = 0.05MPa) shown in Fig. (**2**). The minimum void ratio was established from the experimental results reported by Hoffmann, Alonso and Romero [144] where a dry density of 2.05 g/cm^3 was obtained at a vertical stress of 100 MPa.

Romero, Gens and Lloret [109] also reported the results of an expansion pressure test at constant volume performed in a suction controlled oedometer cell using a sample with an initial suction of 1.8 MPa and dry volumetric weight of 13.7kN/m^3. The test started from the as-compacted condition and suction was reduced by steps up to saturation. Then a drying-wetting cycle was applied. Between each step, enough time was given for suction to equilibrate. Only the vertical stress was measured during the tests. The experimental results along with the numerical simulation are shown in Fig. (**11a**). Here only the initial wetting and subsequent drying path are presented. As aforementioned, depending on the initial stress state of the soil, these tests may show two main stages. In the first stage, the mean net stress increases to equilibrate the tendency of the soil to swell as suction reduces. In the second stage, the stress path follows the LCYS up to saturation. This behavior can be observed in Fig. (**11b**) where the as-compacted LCYS and wetting-drying paths have been plotted in the axes of mean effective stress *versus* suction. A third stage may occur if drying-wetting cycles are performed after complete saturation of the sample. During this stage, the stress path followed by the sample comes back along the LCYS (when plotted in the effective stress axis) because no hardening is experience during stage 2.

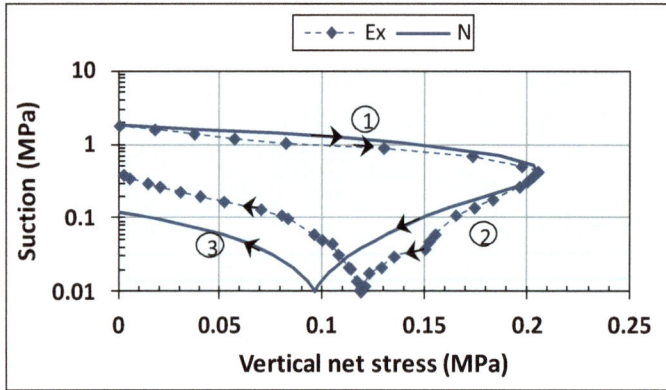

(a)

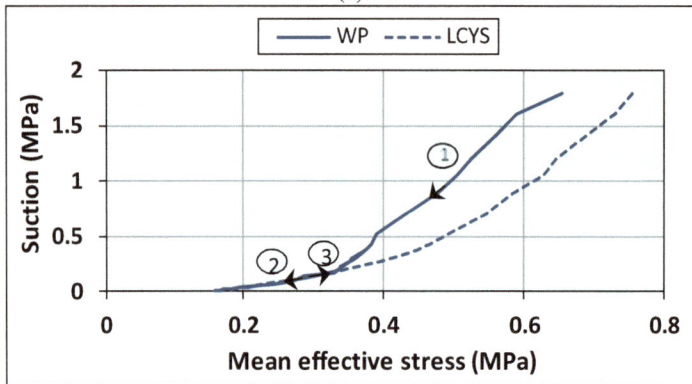

(b)

Fig. (11). Expansion pressure test under null volume change conditions. (**a**) Experimental and numerical vertical net stress against suction and (**b**) initial LCYS and wetting-drying path. Experimental data from [109].

The numerical simulation of this test was obtained from Eqs. (10.3) and (10.4) for stage 1 and Eq. (10.5) for stages 2 and 3. The drying branch of the SWRCs presented in Fig. (**5a**) has been used to obtain the LCYS. This yield surface represents the null volumetric strain path (*i.e.*, neutral loading) and thus the natural path for constant volume tests. The WP followed by the sample takes into account the change in mean net stress as suction reduces and the soil tends to expand. When the WP meets the LCYS, the tendency of the sample to expand arrests, the volume of the sample remains constant, suction decreases and the mean net stress adjust its value to follow the LCYS. This last phenomenon can be explained as follows: when the soil is wetted the additional contact stress between particles generated by water menisci reduces, the material becomes less stiff, more deformable and prone to collapse. Therefore, in order to avoid this collapse and keep the volume of the sample constant, the net stress decreases. The results of this kind of test are very sensitive to the shape of the retention curves of the soil. Some differences between numerical and experimental mean net stresses (expansion pressure) are observed in Fig. (**11a**) although the shapes of the curves are quite similar.

If the WP does not cross the LCYS, the second stage of the test does not occur and the mean effective stress shows a continuous increase up to saturation. The expansive pressure thus obtained represents the value of the suction stress χs. In addition, an angle of 45° is reached at full saturation as experimentally observed by Blight [138] and Kassiff and Ben Shalom [139].

Alonso, Lloret, Gens and Yang [150] performed a series of tests on samples made of pellets of Boom clay statically compacted at dry densities ranging from 13.0 to 15.0 kN/m³. The pellets were obtained from a sample initially compacted at a dry density of 20 kN/m³ and water content of 3%. The sample was crushed and sieved to obtain aggregates of an average size of 2 mm. Suction controlled oedometer test were performed on samples compacted at a dry density of 14 kN/m³. The SWRC of a similar material with dry density of 13.0 kN/m³ is reported by Alonso, Romero and Hoffmann [133]. The fitting of the SWRC to obtain the parameters of the porous-solid model is shown in Fig. (**12**). During the test program, samples were first loaded to different vertical stresses at constant water content then suction was reduced to a value ranging between 0.7 to 0.45 MPa and finally wetting-drying cycles were applied. Each test lasted from 3 to 4 months. The results of these tests and their numerical simulation are shown in Figs. (**13-17**). Unfortunately, the initial suction for these samples was not reported. As samples were loaded at different vertical stresses, they may show different initial suctions. In any case, the initial suction for all samples was established at 5 MPa. This value was selected by fitting the numerical slope with the experimental value for the first wetting cycle in the axis of suction against volumetric strain.

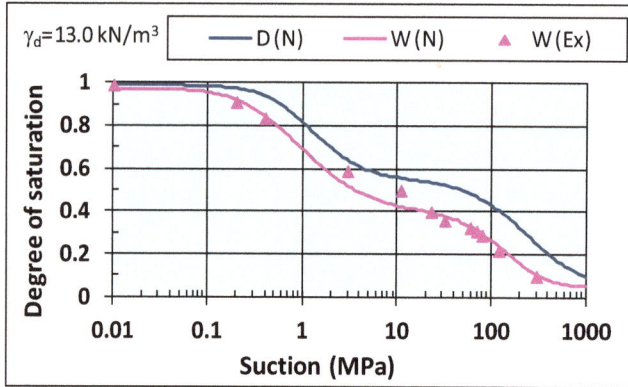

Fig. (12). Experimental SWRC and numerical fitting with the porous-solid model. Experimental data from [133].

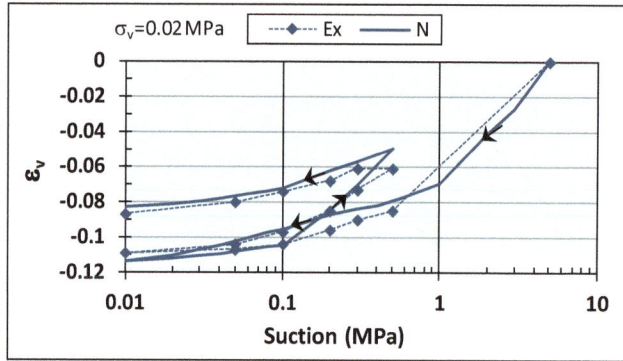

(a)

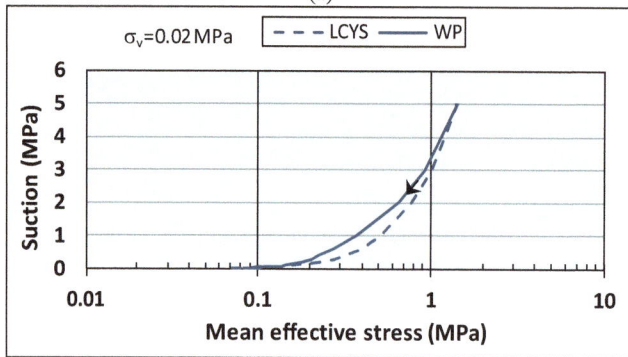

(b)

Fig. (13). (**a**) Volumetric strains during wetting-drying cycles for the sample loaded at 0.02 MPa and (**b**) initial LCYS and WP. Experimental data from [150].

Even if the comparisons between numerical and experimental results show general agreement there are also some inconsistencies. For example: for the sample tested at a vertical stress of 0.1 MPa shown in Fig. (**15a**), the model largely over-predicts both the expansion and subsequent collapse during the first wetting semi-cycle. The same happens with the sample tested at a vertical stress of 0.21 MPa and shown in Fig. (**16a**). In addition, during the following numerical wetting-drying cycles of the former, important volumetric deformations occur when compared with the experimental results. It is possible that, when the unsaturated material is loaded close to the apparent preconsolidation stress, certain length of the WP superposes to the LCYS and, therefore, neutral loading occurs and volumetric strains are null along this portion of the WP. Presently, the model only considers pure expansion (when the WP travels on the left side of the LCYS) or a combination of collapse and expansion (when the WP travels on the right side of the LCYS). Better results could be obtained if neutral loading is included during the simulations. However, this requires the experimental determination and the numerical simulation of the SWRC and LCYS being quite precise in order to correctly capture this phenomenon (compare Figs. **15b** and **16b**).

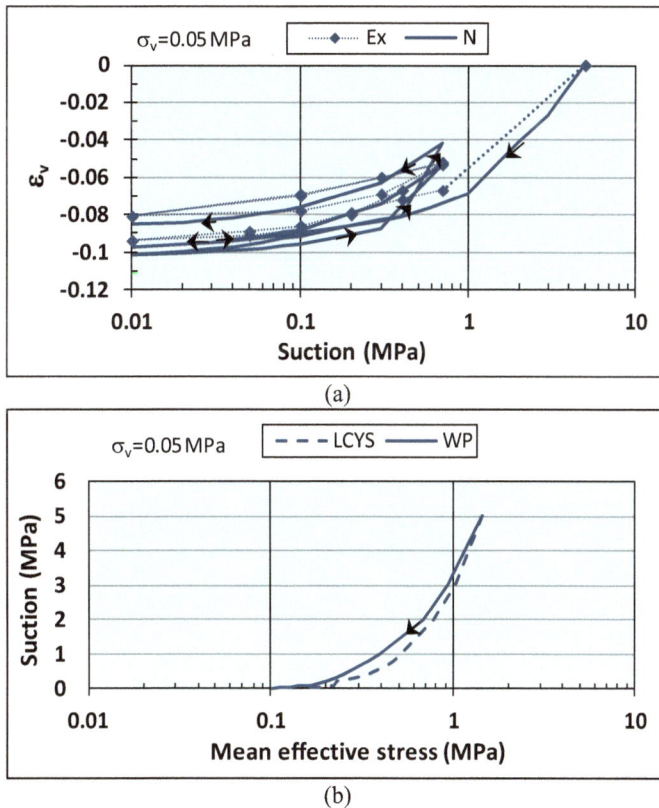

(a)

(b)

Fig. (14). (**a**) Volumetric strains during wetting-drying cycles for the sample loaded at 0.05 MPa and (**b**) initial LCYS and WP. Experimental data from [150].

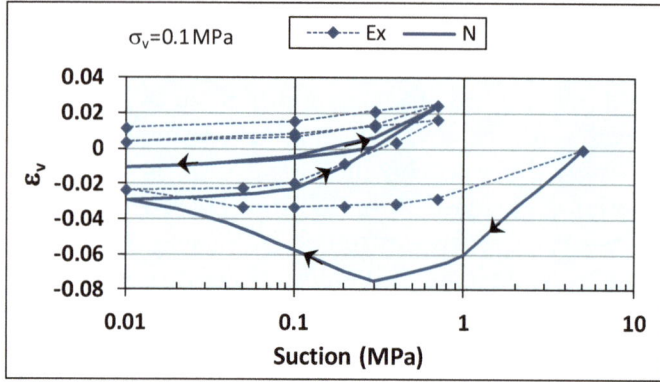

(a)

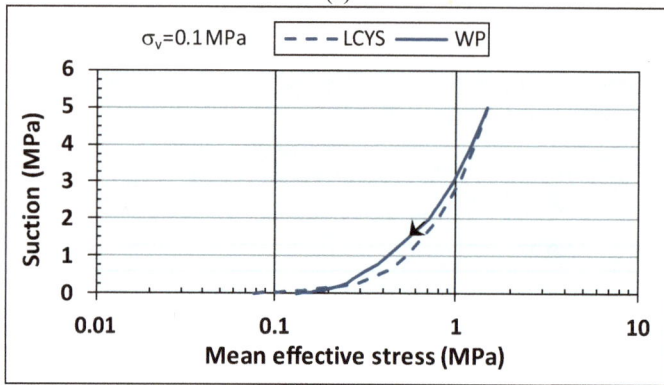

(b)

Fig. (15). (a) Volumetric strains during wetting-drying cycles for the sample loaded at 0.1 MPa and **(b)** initial LCYS and WP. Experimental data from [150].

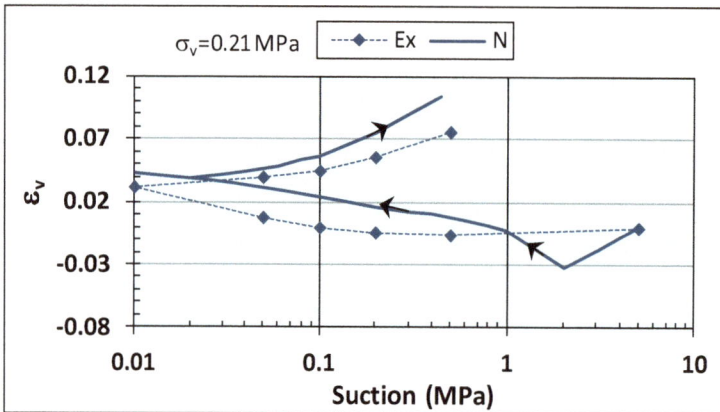

(a)

Fig. 16 cont.....

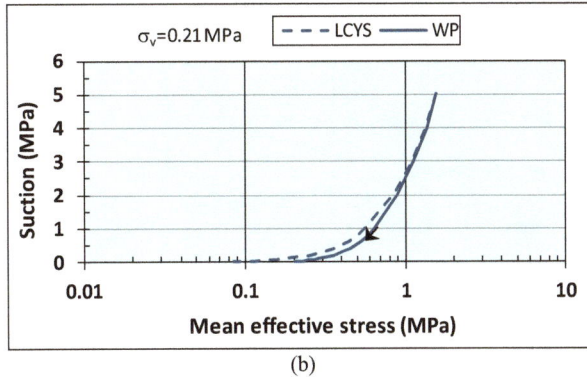

(b)

Fig. (16). (**a**) Volumetric strains during wetting-drying cycles for the sample loaded at 0.21 MPa and (**b**) initial LCYS and WP. Experimental data from [150].

Based on loading-unloading tests reported by the same authors, the values adopted for the compression parameters were λ_e = -0.13 and κ_e = -0.025. For the relative density, the following maximum and minimum void ratios were adopted: e_{max} = 1.35 and e_{min} = 0.30. The maximum void ratio is larger than in the previous tests (Romero, Gens and Lloret [109]) because, even if the soil is the same, the samples used in the present case were made of highly compacted pellets exhibiting large inter-pellet pores (Alonso, Lloret, Gens and Yang [150]).

For practical applications, it may be convenient to use a simpler method to obtain parameter χ. For example, Vanappalli, Fredlund, Pufahl and Clifton [21]; Lu, Godt and Wu [151]; Alonso, Pereira, Vaunat and Olivella [120] among others, have proposed to approximate parameter χ to the effective degree of saturation of the soil, in the form $\chi = S_{we} = (S_w-S_{wr})/ (S_{ws}-S_{wr})$ where S_{wr} and S_{ws} represent the residual degree of saturation and the maximum degree of saturation reached during wetting. Fig. (**18**) shows the results for this case. It can be observed that a major difference with the results shown in Fig. (**17**) is the relative position of the wetting path with respect to the LCYS. While in Fig. (**17b**) the initial wetting path shows an expansion up to 3 MPa in suction, in Fig. (**18b**) the wetting path only shows expansion up to 2 MPa in suction. Another important difference is the magnitude of effective stresses which are larger in Fig. (**18**). This affects the volumetric strains of the sample as shows the comparison between Figs. (**17a and 18a**). Then, when doing the simplification $\chi = S_{we}$ some loose of precision can be expected in the numerical results. This same conclusion was reached when modeling the collapse upon wetting phenomenon (Rojas, Pérez-Rea, López-Lara, Hernández and Horta [152]).

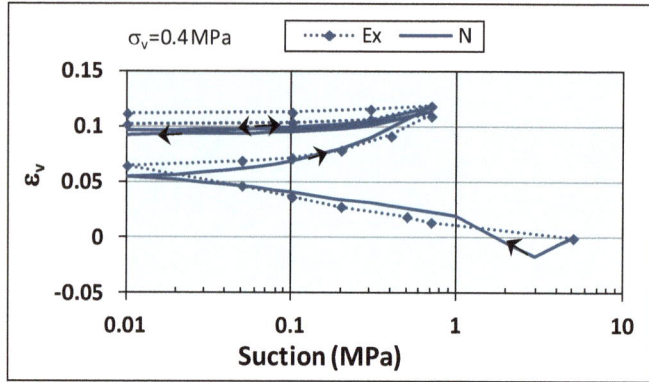

(a)

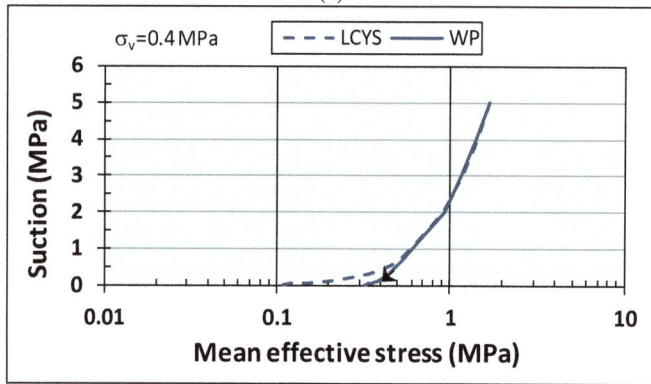

(b)

Fig. (17). (a) Volumetric strains during wetting-drying cycles for the sample loaded at 0.4 MPa and **(b)** initial LCYS and WP. Experimental data from [150].

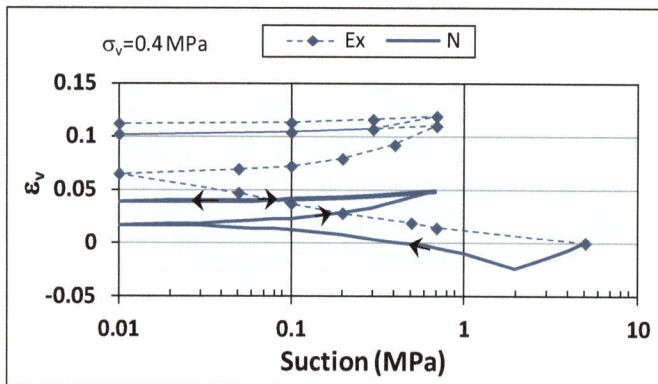

(a)

Fig. 18 cont.....

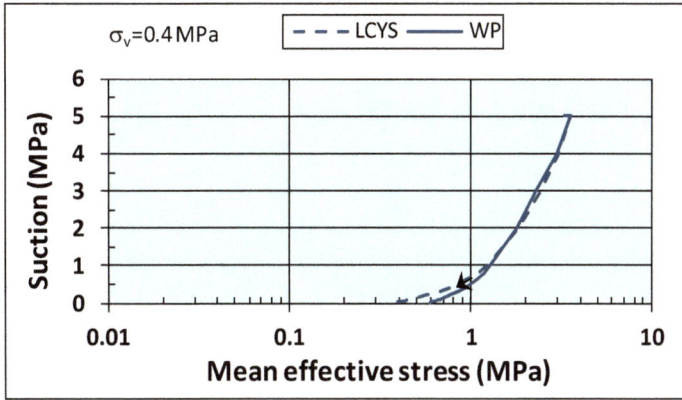

(b)

Fig. (18). Simulation for the case $\chi = S_{we}$. (**a**) Volumetric strains during wetting-drying cycles for sample loaded at 0.4 MPa and (**b**) initial LCYS and WP. Experimental data from [150].

CHAPTER 11

Hydro-Mechanical Coupling

Abstract: The phenomenon of hysteresis during wetting-drying cycles can be simulated by the porous-solid model developed in chapter 3. This model employs the current pore-size distribution of the material. The term "current pore-size distribution" means that the size of pores can be updated as the soil deforms. In that sense, the porous-solid can be used advantageously for the development of fully coupled hydro-mechanical constitutive models as the influence of the volumetric deformation on the retention curves and effective stresses can be easily assessed. This is done by including some experimental observations related to the behavior of the pore size distribution of soils subjected to loading or suction increase. This methodology avoids using any additional parameter or calibration procedure for the hydro-mechanical coupling of unsaturated soils.

Keywords: Effective stresses, Evolution of pore size distribution, Hydro-mechanical coupling, Hysteresis, Loading, Macropores, Mean size of pores, Pore size distribution, Porous-solid model, Soil-water retention curves, Suction increase, Unsaturated soils, Volumetric reduction, Volumetric strain.

11.1. INTRODUCTION

The aim of this chapter is reproduce the shift of the soil-water retention curve (SWRC) generated by plastic volumetric deformations. Doing so, fully coupled constitutive models for unsaturated soils can be developed. To this purpose, the porous-solid model developed in Chapter 4 is used. The idea is to take advantage of the fact that this model is based on the current pore size distribution (PSD) of the material. Then, by doing some assumptions on the way the PSD changes with volumetric deformations, an analytical expression can be derived. Using this procedure, no additional parameters are required for the porous-solid model and therefore no further calibration process is needed.

This paper starts with a brief description of some of the most relevant models used to include the hydro-mechanical coupling in unsaturated soils. Then, the porous-solid model is presented. Afterwards, some assumptions regarding the way in which the PSD changes with the volumetric response of the material are considered and an analytical equation is derived. Later, this proposal is evaluated by doing some numerical and experimental comparisons.

The hydro-mechanical coupling in unsaturated soils was probably first mentioned by Wheeler [145]. This researcher stated that complete constitutive models for unsaturated materials should include information on the water content or degree of saturation. To that purpose, Wheeler [145] proposed the use of the specific water volume (v_w) as the second volumetric state variable. This variable comes in addition to the specific volume ($v = 1 + e$) generally treated as the first volumetric state variable. The specific water volume is defined as

$$v_w = 1 + S_w e$$

where S_w and e represent the degree of saturation and the void ratio of the material, respectively. Wheeler [145] also indicated that fully coupled models should incorporate the hydraulic hysteresis and the effect of the state of stresses on the hydraulic behavior. The hysteresis of the SWRC causes the water content to be dependent not only on the value of suction but also on the wetting-drying path followed by the material. In addition, the volumetric deformation of the soil produces a shift of the SWRC in the axes of suction.

It is presently acknowledged that the best way to develop fully coupled hydro-mechanical models is on the basis of the effective stress principle. In recent years, several effective stress constitutive models that take account of the hysteresis of the SWRC have been developed. Among the most remarkable are the models by Vaunat, Romero and Jommi [90], Jommi [154], Buisson and Wheeler [155], Wheeler, Sharma and Buisson [22], Gallipoli, Gens, Sharma and Vaunat [23], Sheng, Sloan and Gens [25], Tamagnini [24] and Sun, Sheng and Sloan [156]. Different solutions have been proposed by earlier researchers to include the influence of the volumetric deformation on the SWRC (Vaunat, Romero and Jommi [90], Kawai, Kato and Karube [157], Gallipoli, Wheeler and Karstunen [158], Wheeler, Sharma and Buisson [22], Simms and Yanful [79], Koliji, Laloui and Vuillet [88], Sun, Sheng, Cui, Sloan [122], Nuth and Laloui [159], Tarantino [160], Mâsín [161], Sheng and Zhou [162], Gallipoli [163], Salager, Nuth, Ferrari and Laloui [164], Zhou and Ng [140]). Some of these models include the shift of the SWRC dependent on the volumetric deformation ([90, 163, 165]). The diffe-rent solutions adopted by some of these models on this issue are reviewed below.

Vaunat Romero and Jommi [90], adopted the elastoplastic framework of the Barcelona Basic Model (Alonso, Gens and Josa [7]) and included the effect of hysteresis and the state of stresses on the hydraulic behavior of the material. The latter was considered by establishing all possible hydraulic states of the sample in the void ratio-water ratio-suction space. The water ratio was defined as the ratio between the volume of water and the volume of solids. These researchers also considered that macropores are the only responsible for the volumetric deformation of soils. By including the void ratio in a modified van Genuchten formulation of the SWRCs, the influence of the irreversible deformation on the hydraulic behavior of soils is taken into account. For example the water ratio $e_{wD,W}$ at drying (D) or wetting (W) was expressed as

$$e_{wD,W} = e_{wm} + (e - e_{wm})\left[1 + (\alpha_{D,W}s)^{n_{D,W}}\right]^{-m_{D,W}}\left[1 - \frac{\ln\left(1 + \frac{s}{s_{mD,W}}\right)}{2}\right]$$

where e_{wm} is the water ratio of micropores. The values of soil parameters $\alpha_{D,W}$, $s_{mD,w}$, $m_{D,W}$ and $n_{D,W}$ also depend on the path followed by suction: drying (D) or wetting (W). A similar approach to include the influence of void ratio on the SWRC was proposed by Della Vecchia, Jommi and Romero [121].

Khalili, Habte and Zargarbashi [166], simulated the SWRC using a modified Brooks and Corey [167] equation written as a function of the effective degree of saturation (S_{we}), defined as

$$S_{we} = \frac{S_w - S_{wr}}{1 - S_{wr}}$$

were S_{wr} represents the residual degree of saturation. Therefore, the main curves are simulated in terms of the air entry (s_{ae}) or air expulsion (s_{ex}) value for the drying or wetting curve, respectively, in the form

$$S_{we} = \begin{cases} 1 & \text{for} \quad s < s_e \\ \left(\dfrac{s_b}{s}\right)^{\lambda_p} & \text{for} \quad s \geq s_e \end{cases}$$

where s represents the current suction of the material and s_b takes the value of s_{ae} or s_{ex} for the main drying or wetting curve, respectively. Parameter λ_p is called the pore size distribution index and controls the slope of the main retention curves. For the case of wetting-drying cycles, parameter S_{we} depends on the value of suction at the inversion point following a drying (s_{rd}) or wetting (s_{rw}) path and takes the values

$$S_{we} = \begin{cases} \left(\dfrac{s_{ae}}{s_{rd}}\right)^{\lambda_p}\left(\dfrac{s_{rd}}{s}\right)^{\kappa_p} & \text{for drying reversal} \quad s_{rd}\left(\dfrac{s_{ex}}{s_{ae}}\right)^{\frac{\lambda_p}{\lambda_p - \kappa_p}} \leq s \leq s_{rd} \\ \left(\dfrac{s_{ex}}{s_{rw}}\right)^{\lambda_p}\left(\dfrac{s_{rw}}{s}\right)^{\kappa_p} & \text{for wetting reversal} \quad s_{rw} \leq s \leq s_{rw}\left(\dfrac{s_{ae}}{s_{ex}}\right)^{\frac{\lambda_p}{\lambda_p - \kappa_p}} \end{cases}$$

where κ_p represents the slope of the transition between the main drying and main wetting curves after an inversion as represented in Fig. (**1**). The dependency of the SWRC on the volumetric deformation is included by relating the air entry and air expulsion values to the void ratio of the material. In such a case, empirical equations have to be included and model parameters have to be fitted with experimental results. Another disadvantage to the above models is that double structured soils (*i.e.* those showing two main picks in their PSD curve) are not considered.

A more elaborated model was proposed by Zhou and Ng [165]. This model not only includes the effect of the current void ratio in the retention curves but also the influence of the porous structure. This model is an extension of the model developed by Gallipoli [163] and, uses a microstructural state variable defined as $\xi_m = (e_m/e_M + e_m)$ where e_m and e_M represent the micro and macrostructural void ratio, respectively, similar to those proposed by Alonso, Pinyol and Gens [168]. The evolution of the void ratio is established as a function of the mean net stress using a semiempirical equation. The result is an equation of the SWRC written in terms of suction, the initial void ratio and the mean net stress among other parameters.

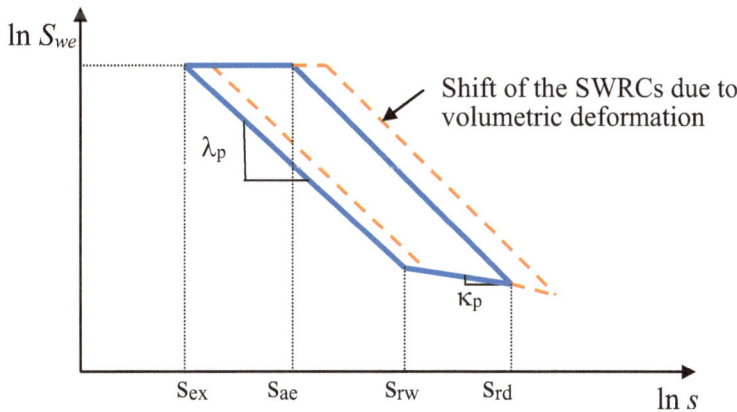

Fig. (1). Model for the SWRCs including hysteresis and volumetric deformation dependency, adapted from [166].

The dependency of the SWRCs on the PSD and the capillary phenomenon has been demonstrated by Jayanth, Iyer and Singh [169, 170], using an isotherm generator. In that sense, Simms and Yanful [45], proposed a model to simulate the shrinkage of pores during suction increase based on the PSD and with this data derive the current SWRC. To model volume change, Simms and Yanful idealized pores as elastic cylinders. Their experimental work included the study of the evolution of the PSD of samples subject to suction increase. These researchers observed that the relative volume of larger pores (macropores) reduced while that of smaller pores (mesopores) relatively increased (Fig. **1** in Chapter 3). By analyzing these results, these authors delimited the range of pores sizes that reduce or increase their relative volume and proposed a model for the redistribution of this volume with suction. Their model also takes account of the entrapment of large pores with smaller ones. This entrapment was included by defining the probability of a pore of certain size to be connected to a smaller one. In this way, modified PSDs were generated to simulate the evolution of the SWRC. Later, based on these principles, Simms and Yanful [79], proposed a 2D deformable pore-network model. The pores were simulated by circles and their sizes were obtained from the PSD of the material. Mercury intrusion porosimetry (MIP) tests were used to define the PSD of the soil. If MIP tests were not available, the PSD was generated by fitting the analytical with the experimental SWRC. A regular 2D network was used for the model where, at each node of the grid, a pore of certain size was randomly assigned. By applying the Young-Laplace equation (Dullien [55]) it was possible to simulate the process of wetting or drying of the soil and reproduce one of the branches of the SWRC. Pores were able to shrink when load or suction increased according to the following relationship

$$\Delta r = (\Delta p'/C)\, r$$

where r and Δr represent the initial pore radius and its decrement, $\Delta p'$ is the effective stress increment and C (1/kPa) is a stiffness parameter of the material. According to the above equation, larger pores will shrink more than smaller ones. The SWRC at drying or wetting could be obtained by adding all water drained or suck up by pores for different values of suction. Isolated pores were also included in the model leading to the simulation of a residual degree of saturation. Unfortunately, this model is unable to simulate properly the phenomenon of hysteresis of the SWRC as to that purpose two different types of elements (with their own pore size distribution) are required (Dullien [55]). Although, single type element models can simulate some kind of hysteresis, they are unable to independently control the positions and shape of the drying and wetting branch.

11.2. PROCEDURE

As stated before, when a compacted soil is subjected to isotropic loading or suction increase, macropores reduce their volume while that of mesopores relatively increases (Simms and Yanful [45, 78, 79]). Koliji Laloui, Cusinier and Vulliet [88] also observed the same phenomena during suction increase although, these authors, reported that the largest macropores showed no collapse. In contrast, Thom, Sivakumar and Sivakumar [84] observed that both macropores and mesopores reduced their volume, although, the reduction in the volume of macropores was much larger than for mesopores. Thom Sivakumar and Sivakumar [84] also reported that the mean size of macropores continuously reduces as the soil compresses. Finally, these authors also reported that new sizes of pores (their range between those of macropores and mesopores) appeared during the compression of the sample. According to these experimental results, it seems that the evolution of the PSD during the soil deformation greatly depends on the type of material. Under these circumstances, the evolution of the PSD of a certain soil can be simulated by the porous-solid model in three different ways. In one case, it can be assumed that during loading or suction increase, a certain amount of macropores collapse reaching a different size including that of mesopores. Other assumption is that all macropores reduce continuously in size as the material compresses. Finally, a combination of these two assumptions can be considered.

As a first approximation to the evolution of the PSD of soils during compression, the first assumption is considered herein in the following way: when the soil compresses, macropores shrink to the size of mesopores. This is equivalent to consider that the volume of macropores reduces while that of mesopores relatively increases. It is also considered here that macropores and mesopores change in volume homogeneously, in other words, all pore sizes change in the same proportion as proposed by Simms and Yanful [79]. This assumption is experimentally confirmed by the results shown in Fig. (1), Chapter 3. This behavior can be easily introduced in the porous-solid model when the PSD is plotted in the axis of relative volume against pore size. In such a case, it is easy to reduce the relative volume of macropores while that of mesopores automatically increases. This phenomenon is controlled by the relative volume factor for macropores (R_{vM}) referred before. This factor reduces the total number of macropores of each size (N_{TMi}) to a number (N_{rMi}) compatible with the structure of the real material. The total number of macropores of each size (N_{TMi}) is obtained by considering that all nodes of the network are occupied by macropores. Therefore, the real number of macropores of each size is given by

$$N_{rMi} = R_{vM} N_{TMi}$$

Then, the real volume of macropores (V_{rM}) in a 3D porous-solid model is

$$V_{rM} = N_{rMi} \sum_{1}^{n} \frac{4}{3} \pi r_i^3 \tag{11.1}$$

where n represents the number of ranges of macropores, r_i is the mean radius of macropores of range *i*. This volume is added to the volume of mesopores to obtain the total volume of cavities. Because experimental results show that both the wetting and the drying retention curves shift during volumetric deformation this means that both the largest cavities and the largest bonds reduce in size. Therefore, a similar relationship to Eq. (11.1) has to be used to reduce the volume of large bonds (V_{rB}). The addition of the total volume of cavities and the total volume of bonds represents the volume of voids of the sample. It is considered here that only volumetric plastic strains produce the shrinkage of large pores and thus the shift of the SWRCs. Therefore, if a material shows a plastic reduction of the volume of voids (ΔV_v^p) due to the shrinkage of macropores and large bonds, it results in the following relationship

$$\Delta V_v^p = \Delta e^p V_s = (\Delta e_M^p + \Delta e_B^p)V_s = \Delta e_M^p \left(1 + \frac{\bar{R}_B}{\bar{R}_M}\right)V_s = V_{rM}\left(1 - \frac{R_{vM1}}{R_{vM0}}\right) + V_{rB}\left(1 - \frac{R_{vB1}}{R_{vB0}}\right) \quad \textbf{(11.2)}$$

where Δe^p represents the plastic reduction of void ratio and V_s is the volume of solids, Δe_M^p and Δe_B^p are the plastic reduction of void ratio produced by the shrinkage of macropores and large bonds, respectively, $\bar{R}_M$ and $\bar{R}_B$ represent the mean size of macropores and large bonds, respectively, while R_{vMo}, R_{vM1}, R_{vBo} and R_{vB1} are the initial and current relative volume factor for macropores and bonds, respectively. In the above equation it has been considered that the reduction in volume of large bonds can be obtained by multiplying the volume reduction of macropores by the ratio between the mean size of large bonds and macropores. From the above equation, the value of R_{vM1} and R_{vB1} can be obtained

$$R_{vM1} = R_{vM0}\left[1 - \frac{\Delta e^p V_s}{V_{rM}}\left(\frac{\bar{R}_M}{\bar{R}_M + \bar{R}_B}\right)\right] \quad \textbf{(11.3)}$$

$$R_{vB1} = R_{vB0}\left[1 - \frac{\Delta e^p V_s}{V_{rB}}\left(\frac{\bar{R}_B}{\bar{R}_M + \bar{R}_B}\right)\right] \quad \textbf{(11.4)}$$

In this way, the procedure to simulate the evolution of the SWRC as the soil deforms, requires the following steps: first, the initial experimental SWRCs are fitted with the porous-solid model. In this process, an initially proposed PSD is successively modified until the best fit for both curves is obtained. The final result is the initial distribution of the size of pores. In the same way, the experimental grain size distribution (GSD) has to be fitted with the analytical curve to obtain the size distribution of solids. This allows building the porous-solid model. Then parameters λ_e and κ_e are established from experimental results. This allows simulating the volumetric behavior of the soils at loading and unloading. Then, by using Eqs. (11.3) and (11.4) it is possible to determine the evolution of the PSD of the material dependent on the volumetric plastic deformation. Finally, for each increment of load, a PSD is obtained, and with the use of the porous-solid model, it is possible to determine the progression of the SWRC as the soil deforms.

11.3. NUMERICAL AND EXPERIMENTAL COMPARISONS

11.3.1. Tests by Chiu and Ng [146]

Chiu and Ng [171], performed a series of tests on a completely decomposed tuff classified as a low plasticity silt (ML). This material shows fractions of sand, silt and clay of 24%, 71% and 5%, respectively. Compacted soil specimens 19 mm in height and 70 mm in diameter were prepared at a dry density of 15.1 kN/m³ and a water content of 16%. The maximum dry density for this material obtained in the standard Proctor tests was 17.6 kN/m3 for an optimum water content of 16.3%. After compaction, soil specimens were saturated with deaired water during 48 h. These samples were placed in a suction controlled double-wall triaxial apparatus to obtain the SWRCs. The axis translation technique was used to control suction in the soil sample. This equipment allowed accurate measurements of the volume change of the sample during the test.

The experimental tests included the SWRC of three isotropically loaded samples at three different confining stresses. The SWRC was determined during a large drying-wetting cycle. The applied isotropic stress and the specific volume before and after drying for each sample are shown in Table **1**. The procedure to obtain the SWRC consisted in three stages: first, the suction equalization stage; second, the loading stage, and third, the drying–wetting cycle to obtain the SWRC. An initial suction of 0.0001 MPa was applied to the specimens by means of an outlet tube. The suction equalization was terminated when the water flow was lower than 0.1 cm³ per day.

Table 1. Soil samples of decomposed tuff tested in the triaxial cell [171].

Test	Isotropic Stress (MPa)	1+e Before Drying	1+e Before Wetting
I-0	0	1.795	1.689
I-40	0.04	1.736	1.662
I-80	0.08	1.717	1.653

Fig. (**2a**) shows the gradation curve of the soil along with the numerical fitting. From this fitting it is possible to obtain the mean size and standard deviation of solid particles. The SWRCs of the samples tested at different isotropic stresses are presented in Fig. (**2b**). Fig. (**2c**) shows the best fit for the SWRCs of the unconfined sample. From this fitting, the PSD shown in Fig. (**2d**) is obtained. In this same figure the GSD of the soil obtained from the fitting of the gradation curve shown in Fig. (**2a**) is presented. Unfortunately, the experimental PSD of the soil was not reported. Table **2** shows the values of the porous-solid model parameters describing the GSD and PSD shown in Fig. (**2d**).

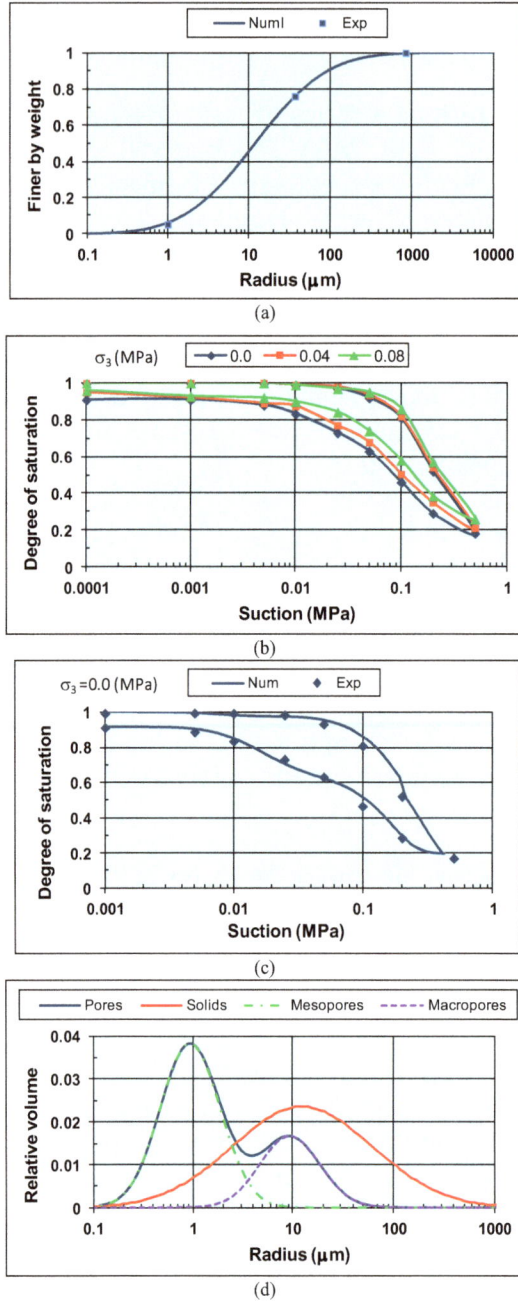

(a)

(b)

(c)

(d)

Fig. (2). Decomposed tuff: (**a**) gradation curve and analytical fitting; (**b**) retention curves at different isotropic stresses; (**c**) best fit for the SWRCs of the unloaded sample; (**d**) analytical PSD and GSD. Experimental data from [171].

Table 2. Parameters of the porous-solid model for the decomposed tuff.

Parameter	S_1	S_2	B_1	B_2	Sol
$\overline{R}$	0.35	3.5	0.07	0.6	0.07
δ	2.0	2.0	2.8	2.9	5
R_v	.0055		.01		
F_s					0.715

Note: S_1 = mesopores; S_2 = macropores; B_1 and B_2 = bonds; Sol = solids; $\overline{R}$ = mean size; δ = standard deviation; R_v = relative volume factor (pores); F_s = shape factor (solids).

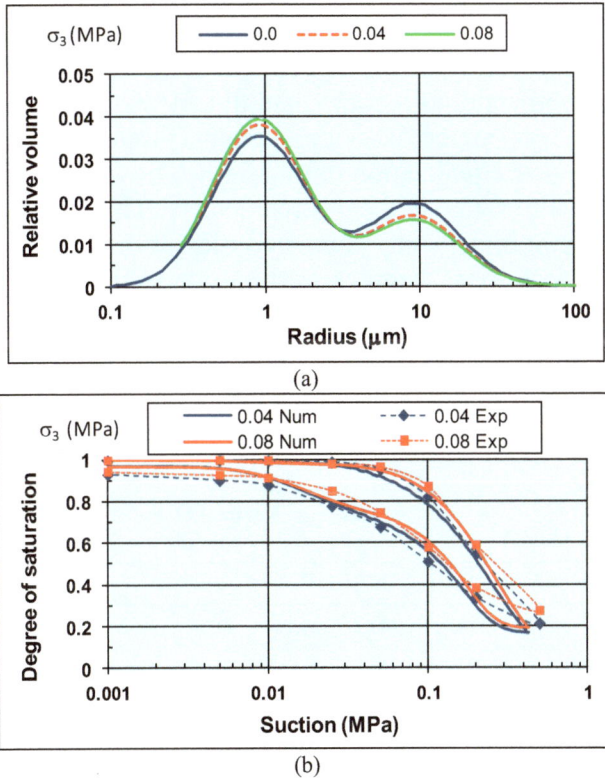

(a)

(b)

Fig. (3). (a) Evolution of the PSD with isotropic stress; **(b)** SWRCs for samples confined at 0.04 MPa and 0.08 MPa. Experimental data from [171].

Once the PSD and GSD of the soil have been defined, the porous-solid model can be build. Then, using Eqs. (11.3) and (11.4), it is possible to obtain the evolution of the PSD with the applied stress as shown in Fig. (**3a**). This figure shows the reduction of the relative volume of macropores and the increase in mesopores with the confining stress. Finally, with the PSD for each confining stress it is possible to simulate the SWRCs for each sample. These curves are shown in

Fig. (**3b**) along with the experimental points for isotropic stresses 0.04 and 0.08 MPa. In this figure, it can be observed that numerical results follow quite close the experimental points.

11.3.2. Tests by Ng and Pang [147]

Another series of tests were performed by Ng and Pang [172]. These authors tested a completely decomposed ash tuff classified as highly compressive silt (MH) with 4.9% gravel, 20.1% sand, 36.6% silt and 37.1% clay. The maximum dry density of this material was 16.03 kN/m^3 for an optimum water content of 22%. The SWRCs of the soil samples were obtained with a modified pressure-plate extractor. With this equipment, vertical loads were applied on samples placed inside an oedometer ring. The specimens used to obtain the SWRCs were prepared by static compaction. First, the natural soil was dried at 45° C during 48 h. Then the material was pulverized with a hammer, water was added and keep in plastic bags for moisture equalization during 24 h. Then the soil was placed in an oedometric cell and compacted to the desired density. The specimens were 70 mm in diameter and 20 mm high. Some characteristics of the compacted samples are shown in Table 3. Later, these samples were saturated and preconsolidated to the required vertical stress. Then, they were subject to a drying-wetting cycle where the vertical deformations along with the flow of water were registered. Using these results, the SWRCs were obtained.

Table 3. Some characteristics of the ash tuff (Ng and Pang [172]).

Test	γ_{di} (kN/m^3)	e_i	w_i (%)	Vertical Stress (MPa)	Max. Suction (MPa)	e_f
CDVR1	14.7	0.782	30.3	0	0.4	0.758
CDVR2	14.7	0.782	30.3	0.04	0.2	0.724
CDVR3	14.7	0.782	30.3	0.08	0.2	0.701

Note: γ_{di} = initial dry density; e_i, e_f = initial and final void ratio, respectively; w_i (%) = initial water content.

The experimental gradation curve for this material is shown in Fig. (**4a**) along with the fitted analytical curve. Fig. (**4b**) shows the fitting of the SWRCs for the unloaded sample. From this fitting, the PSD shown in Fig. (**4c**) was obtained. This same figure shows the GSD obtained by the fitting of the gradation curve shown in Fig. (**4a**). Table **4** presents the parameters of the model derived from this process.

Once the parameters of the model have been defined it is now possible to simulate the evolution of the PSD with the vertical stress as shown in Fig. (**5a**). Again, it can be observed that macropores reduce their volume while mesopores increase their, with the vertical stress. Finally, Fig. (**5b**) shows the analytical and experimental SWRCs for samples loaded at 0.04 and 0.08 MPa. The analytical curves were obtained using Eqs. (11.3) and (11.4) and includes the change in void ratio produced by loading as indicated in Table **3**. As it can be observed in this figure, the porous-solid model reproduces fairly well the evolution of the SWRCs due to volumetric deformation.

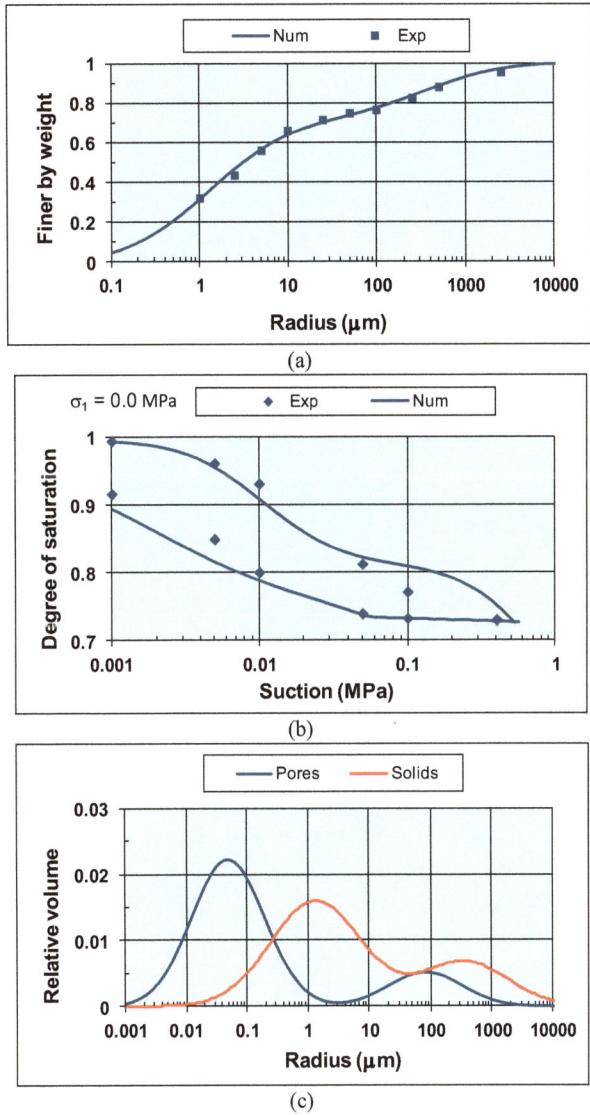

(a)

(b)

(c)

Fig. (4). Results of the ash tuff: (**a**) gradation curve and numerical fitting; (**b**) best fit for the SWRCs of the unloaded sample; (c) analytical PSD and GSD. Experimental data from [172].

Table 4. Parameters of the porous-solid model for the ash tuff.

Parameter	S_1	S_2	B_1	B_2	Sol_1	Sol_2
$\overline{R}$	0.001	1.7	0.0005	0.7	0.004	2
δ	4.0	4.0	4.0	4.0	5	5
R_v	.00000008		.0025		0.000003	
F_s					0.01	

Note: S_1 = mesopores; S_2 = macropores; B_1 and B_2 = bonds; Sol = solids; $\overline{R}$ = mean size; δ = standard deviation; R_v = relative volume factor (pores); F_s = shape factor (solids).

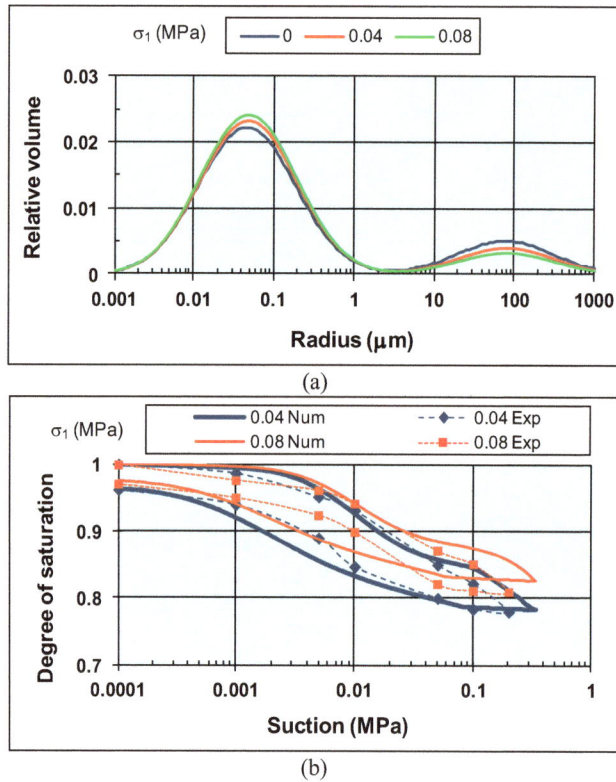

(a)

(b)

Fig. (5). (a) Evolution of the PSD with vertical stress; **(b)** SWRCs for σ_1=0.04 MPa and 0.08 MPa. Experimental data from [172].

11.3.3. Tests by Sun, Sheng, Cui and Sloan [122]

In the previous tests, the SWRCs were obtained after loading the soil samples. A different procedure was used by Sun *et al.* 2007a. In this case, the SWRCs were obtained during the deformation of the sample. These authors used statically compacted samples of Pearl clay at a water content dry of optimum (26%). This material is constituted by 50% silt and 50% clay. The gradation curve of this material is shown in Fig. (**6a**). The samples were compacted in five layers up to a specific vertical stress. The initial voids ratio and degree of saturation of samples was controlled by the compaction energy. The tests were performed in a suction controlled triaxial cell equipped with three radial displacement rings to determine the volumetric deformation of the sample.

One of the tests performed by Sun, Sheng, Cui and Sloan [122] was made on a sample compacted at an initial void ratio of 1.78, then subjected to an isotropic stress of 0.02 MPa with an initial suction of 0.1 MPa. Finally, the sample was subjected to drying-wetting cycles with suction varying from 0.5 to 0.005 MPa. The experimental volumetric behavior of this sample is shown in Fig. (**6b**). Observe that during the first drying and the initial part of the first wetting the sample shows practically no volume change. Then, when suction reduces from 0.06 to 0.005 MPa during the first wetting, the sample collapses. This means that the excerpts of the retention curves obtained from the initial drying and the first part of the wetting up to a suction of 0.06 MPa, represent the initial SWRCs of the soil as depicted in Fig. (**6c**). Therefore, by fitting these points with the analytical curves, the initial PSD of the material can be obtained as shown in Fig. (**6d**). The parameters of the model determined from the fitting process are shown in Table **5**. The analytical volumetric deformation of the sample shown in Fig. (**6b**) was obtained from the elastoplastic framework described by Rojas and Chávez [173] and Rojas, Pérez-Rea, López-Lara, Hernández and Horta [149]. This simulation allowed computing the volumetric strain during the wetting-drying cycles applied to the soil and using Eqs. (11.2) and (11.3), compute the relative volume factor for macropores and large pores as the soil deforms. Fig. (**6d**) also shows the final PSD reached at the end of the test. Observe that in this case the reduction in the volume of macropores and the corresponding increase in mesopores are quite noticeable when compared with the previous results. This is so because the void ratio of the sample during collapse changes from 1.7 to 1.2 (see Fig. **6b**).

Table 5. Parameters of the porous-solid model for Pearl clay, e = 1.78.

Parameter	S_1	S_2	B_1	B_2	Sol_1	Sol_2
$\overline{R}$	0.08	3.6	0.1	1.0	0.1	0.5
δ	3.0	2.0	2.5	3.0	2.2	2.5
R_v	.0025		.025		0.15	
F_s					0.372	

Note: S_1 = mesopores; S_2 = macropores; B_1 and B_2 = bonds; Sol = solids; $\overline{R}$ = mean size; δ = standard deviation; R_v = relative volume factor (pores); F_s = shape factor (solids).

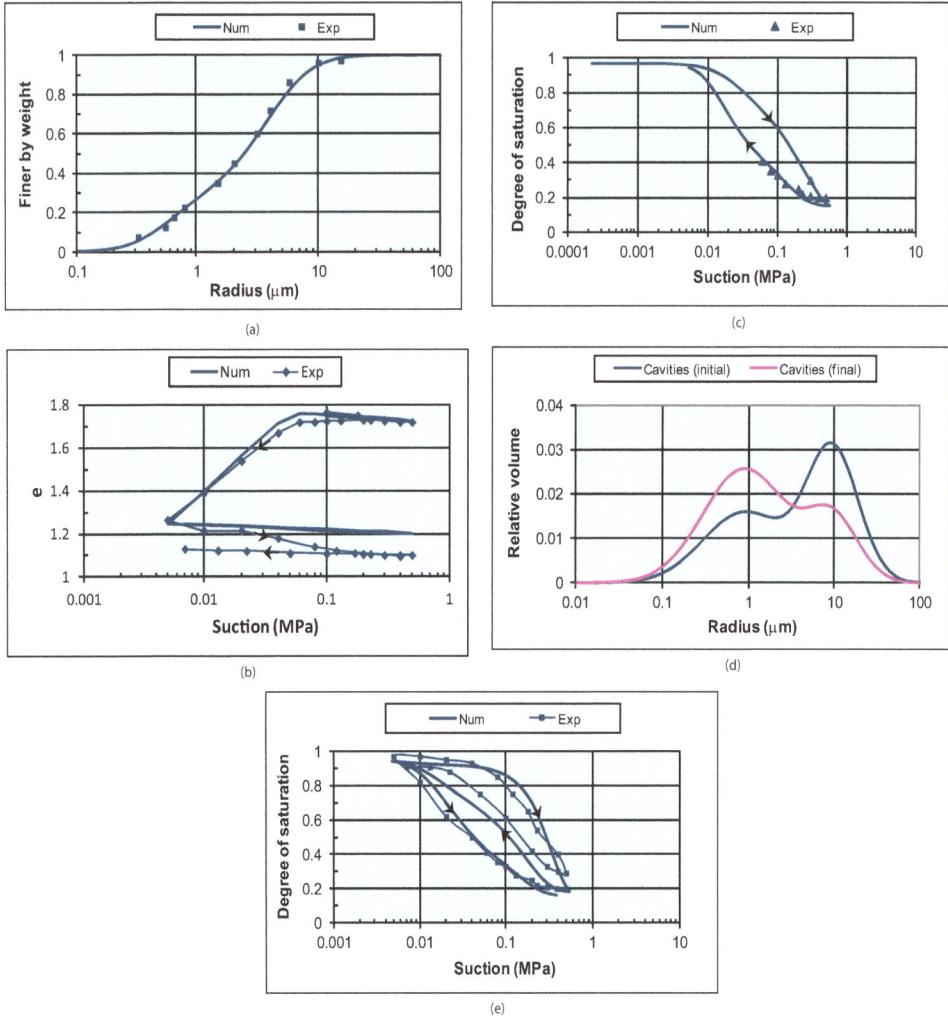

Fig. (6). Pearl clay: (**a**) Gradation curve; (**b**) volumetric behavior of soil; (**c**) initial SWRCs (**d**) initial and final PSD for cavities and bonds; (**e**) shift of the SWRC with suction. Experimental data from [122] and [101].

Whenever the sample shows volumetric plastic deformations, the PSD of the material is updated using Eqs. (11.3) and (11.4). Then, the volume of saturated pores is recalculated in order to obtain the current degree of saturation of the sample. The current volume of saturated pores calculated for the SWRCs are obtained by simulating the wetting or drying process from the initial point or from an inversion point every time that the PSD changes. This means that, for each increment in the volumetric plastic strain, the porous–solid model is restarted from the initial point or from the point where an inversion in suction has occurred and the SWRC is recalculated from this point up to the current suction. Finally, Fig. (**6e**) shows the evolution of the retention curve from the first wetting to the end of the test. It can be observed that the porous-solid model correctly simulates the shift of the SWRCs produced by volumetric deformations.

It is worth noting that the first wetting of the sample does not show any important shift of the SWRC even if the soil collapses during this cycle. It is not until the next drying cycle when the SWRCs show an important shift. This happens because during the initial part of the first wetting there is no volumetric deformation and the PSD remains unchanged. Only when suction reduces farther than 0.06 MPa, the PSD modifies and the remainder of the wetting SWRC slightly shifts. Also notice that during the first wetting and for suctions in the range between 0.06 and 0.005 MPa, each point of the SWRC has been obtained from a different PSD. The process ends with the final PSD shown in Fig. (**6d**).

Although, experimental results are fairly well simulated considering the assumption that macropores collapse to the size of mesopores as the soil deforms, other assumptions can be explored in order to improve these results. However, an analytical expression for the assumption that the mean size of macropores progressively reduces as the soil compresses, still imposes some mathematical problems that have yet not been solved.

CHAPTER 12

A Fully Coupled Model

Abstract: In previous chapters, it has been shown that the principle of effective stresses can be applied to the shear strength, the tensile strength and the volumetric behavior of unsaturated soils. This chapter shows that the critical state line for unsaturated soils shifts with respect to the saturated critical sate line in a quantity that depends on the suction stress. Taking into account this phenomenon and the influence of hydro-mechanical coupling on the behavior of unsaturated soils, a fully coupled general constitutive models for soils is developed. This model is based on the modified Cam-Clay model and includes a yield surface with anisotropic hardening that takes into account the shift of the critical state line with suction. The result is a very simple model with a symmetric stiffness matrix that can be used for saturated, unsaturated and compacted materials.

Keywords: Constitutive models, Critical state concept, Critical void ratio, Effective stress, Elastic zone, Elastoplastic framework, Failure surface, Plastic deformations, Preconsolidation stress, Suction hardening, Tensile strength, Virgin consolidation line, Volumetric behavior, Yield surface.

12.1. BACKGROUND

In the last fifteen years, many hydro-mechanical coupled models for unsaturated soils have been developed using the effective stress concept. Some recent examples of models developed to solve different problems can be found in [174 - 178]. Some models use non-normal flow rules requiring a plastic potential function in addition to a yield surface. Others use elaborated hardening coefficients to reproduce the volumetric behavior of unsaturated soils. Some others include fitting parameters for hydro-mechanical coupling. Models based on the modified Cam-Clay model (MCCM) have been largely used with fair results [25, 121, 179 - 185]. Other approaches have also been used with similar precision, for example hypoplasticity ([186, 187]), generalized plasticity [188], hydromechanical energy dissipation [189], bonding fabric ([99, 190]), among others. The hydraulic behavior is generally included employing the simple elasto-plastic model proposed by Wheeler [22]. Others use the van Genuchten [61] or the Fredlund, Wilson and Fredlund [191] equations for single or double porosity to simulate the main wetting and drying curves. Then, with the aid of some analytical relationships, wetting-drying cycles can be simulated.

Some of the most representative models are those proposed by Russell and Khalili [192], Zhou and Sheng [162] and Ma, Wei, Wei and Li [185]. The model by Russell and Khalili [192] uses Bishop's effective stress equation with Bishop's parameter dependent solely on suction. This model uses the bounding surface concept to generate a smooth transition between elastic and elastoplastic behavior. The loading and the bounding surface show the same shape and are represented by a logarithmic function with isotropic hardening dependent on the volumetric strain. A simple radial mapping rule with center in the origin, is used to define the projection of the state of stresses on the bounding surface. Additionally, a non-associated flow rule is used with a plastic potential obtained from the integration of a dilation rule. The flow rule is dependent on the position of the stress state with respect to the critical state line (CSL) as well as the direction of the normal vector of the plastic potential. This feature ensures contraction or dilatation when the stress state is located below or above the CSL, respectively. The hardening modulus is split in two parts, one for the bounding surface and the other dependent on the distance between the loading and the bounding surface. This last hardening modulus can be defined arbitrarily provided that it becomes null at the bounding surface and uses a parameter which depends on the initial conditions. The model includes the effect of particle crushing for the case of sands tested at large mean stresses. It also considers the shift of both the CSL and the isotropic consolidation line (VCL) due to suction hardening by means of an analytical equation written in terms of suction and the volumetric plastic strain of the sample. This model requires several fitting parameters and functions such as: the exponent of the yield surface $(1/N)$, an initial state or changing conditions parameter k_m, a function defining energy dissipation k_d, a function defining the slope of the isotropic compression line dependent on suction $\lambda(s)$ and a function defining the shift of the VCL with suction. In addition, as the plastic potential does not cross the mean effective stress axis at straight angle, deviatoric strains appear during isotropic loading.

The model by Zhou and Sheng [184] uses Bishop's equation and the effective degree of saturation as Bishop's parameter. Contrary to common models where suction is used as the third axis in addition to mean effective stress and deviator stress, this model uses the effective degree of saturation as the third axis. It employs a bounding and a subloading surface as well as a unified hardening parameter in addition to a hydraulic model to build a coupled hydro-mechanical model. It includes a loading collapse yield surface (LCYS) similar in shape to the Barcelona Basic Model but written in terms of the effective degree of saturation. The compression index also depends on the effective degree of saturation of the soil. A common isotropic hardening rule is used for the yield surface. A modified van Genuchten equation, which includes the influence of volume change on the soil-water retention curves (SWRCs), is employed for the hydraulic model. Wetting-drying cycles can be simulated using the Sheng and Zhou [162] hydraulic model. The initial void ratio is considered as a key variable for the behavior of the soil. It requires two fitting parameters for hydro-mechanical coupling. The phenomena of suction hardening and the shift of the VCL and CSL with suction are not included directly into the model. This model requires 13 parameters.

The model by Ma, Wei, Wei and Li [185] uses Bishop's equation and the degree of saturation as Bishop's parameter to compute effective stresses. This model is based on the elliptic yield surface of the MCCM and couples hydraulic and mechanical behavior. The Feng and Fredlund [193] hydraulic model is used to simulate wetting-drying cycles and includes the dependency of the SWRCs on the plastic volumetric strains. The constitutive model uses a non-associated flow rule through a dilatancy term. Hardening rule considers the effect of volumetric and deviatoric plastic strains. The influence of deviatoric plastic strains is included by means of a parameter dependent on the degree of saturation and suction. It also considers a correction function which accounts for the hardening effect of a non saturated material which depends on the current value of several parameters such as suction, the degree of saturation and the plastic volumetric strain of the sample. This model does not consider the shift of the VCL or CSL due to suction hardening.

This chapter shows that constitutive models based on the Critical State theory for saturated soils can be easily adapted as fully coupled models for unsaturated soils when the phenomena of suction hardening, hydraulic hysteresis and dependency of SWRCs on plastic volumetric strains are properly considered. Specifically, this chapter shows that the MCCM can properly simulate the behavior of unsaturated soils with minor changes. In this way, very simple fully coupled constitutive models for soils can be generated with similar precision to other models.

12.2. CRITICAL STATE

One issue that requires reviewing is the critical state concept for soils tested at different suctions. Wheeler and Sivakumar [83] performed a series of triaxial tests on samples of unsaturated compacted speswhite kaolin. The samples were prepared by static compaction in a mould at 25% water content. The tests were conducted in double-walled triaxial cells designed to accurately measure the volume change of the samples during the test. In Chapter 6, the compression strength of unsaturated samples subjected to different suctions was predicted using the concept of effective stress. These simulations showed that a unique

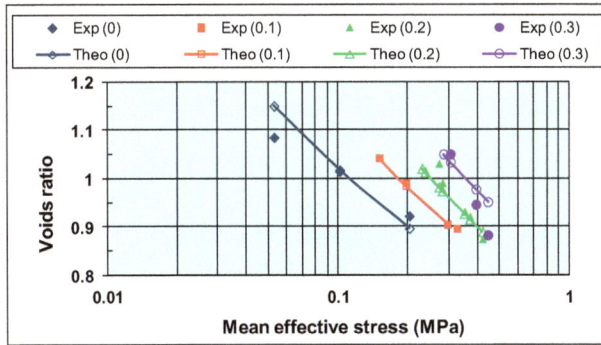

Fig. (1). Critical state for speswhite samples tested at different suctions. Experimental data from [83].

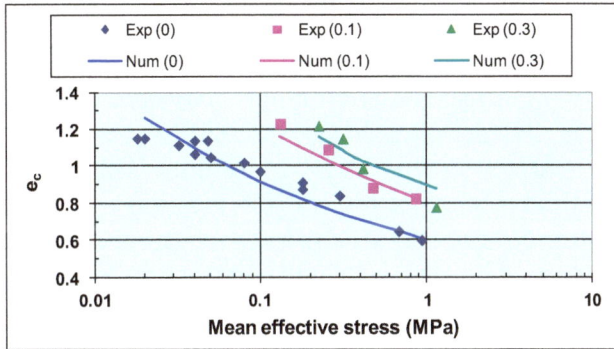

Fig. (2). Critical state for residual gneiss samples tested at different suctions. Experimental data from [92].

failure surface can be obtained when results are plotted on the plane of the mean effective stress against the deviator stress at the critical state. Figs. (**1** and **2**) show the void ratio at the critical state against the logarithm of the effective stress for two different types of soils tested at different suctions. Fig. (**1**) was obtained from data reported by Rojas [87]. These results show that the critical void ratios of samples tested at different suctions align in parallel curves with the slope of the virgin consolidation line (VCL). They also show that the shift of the unsaturated critical sate lines with respect to the saturated critical sate line increases with suction. This same result has been found by other researchers in different soils (see for example [110, 179, 194]). Finally, another important observation is that the shift of the critical state lines corresponds with the shift of the virgin consolidation lines of samples tested at different suctions. This is demonstrated in Figs. (**1** and **2**) since they have been plotted using the χ values obtained from Fig. (**2**) in Chapter 6 and Fig. (**8**) in Chapter 8, respectively. In other words, the shift of the critical state line is produced by the phenomenon of suction hardening and is given by the value of the suction stress. This means that the critical state concept can also be applied to the case of unsaturated materials when the phenomenon of suction hardening is taken into account. These observations open the possibility of developing general constitutive models based on the critical state theory.

12.3. GENERAL ELASTOPLASTIC FRAMEWORK

As established in Chapter 8, when loading is preceded by drying in soil samples, the phenomenon of suction hardening occurs. It is one of the most influencing phenomena when modeling the behavior of unsaturated soils. This phenomenon arises because the components of the effective stress, net stresses and suction stress, act independently one from the other. When a soil dries, the apparent preconsolidation stress increases in a quantity dependent on the increment of suction stress. Consider Fig. (3a) representing the volumetric behavior of the soil in the axes logarithm of the effective mean stress *versus* void ratio. Suppose that a saturated normally consolidated soil is initially subjected to a mean net stress p_{n0} indicated by point 0 in Fig. (3a). If this sample is subjected to drying, the effective mean stress $(p' = p_n + \chi s)$ increases in the quantity χs_0, where s_0 represents the maximum suction reached at drying and χ represents Bishops parameter at this suction. This behavior is sketched in Fig. (3b) in the axes of effective mean stress against suction. During drying, the effective mean stress moves from 0 to A in Fig. (3a). If at this stage, the sample is loaded by an increment of the net mean stress, an initial elastic behavior occurs (path AB) until the apparent preconsolidation stress (p'_0) is reached. This apparent preconsolidation stress is represented by the effective stress at the end of drying added by the suction stress (χs_0) as indicated in Fig. (3b). If the net mean stress increases further, elastoplastic strains occur (dotted path beyond B in Fig. 3a). This means that the increment in the apparent preconsolidation stress after drying (point B) with respect to the initial preconsolidation stress (ps_0) is twice the suction stress χs_0, as indicated in Fig. (3b). Therefore, the LCYS runs parallel to the drying path at a horizontal stress equal to the suction stress χs_0 (see Fig. 3b). In other words, when the effective stress increases in the quantity $\Delta(\chi s)$, the apparent preconsolidation stress increases in the quantity $2\Delta(\chi s)$. This explains why unsaturated soils rapidly show highly preconsolidated behavior when suction increases. In contrast, when the effective stress increases by a net stress increment Δp_n beyond the LCYS, the apparent preconsolidation stress (p'_0) only increases in this same quantity.

The influence of the mean net stress on the suction increase yield surface (SIYS) has been studied by Thu, Rahardjo and Leong [195] who performed a series of triaxial tests on compacted samples of coarse kaolin subjected to different confining stresses. These results show that the SIYS is only slightly affected by the value of the mean net stress. In brief, when suction increases, the LCYS displaces parallel to the drying path in the quantity χs while the SIYS displaces vertically to the maximum value of suction and remains basically horizontal. Accordingly, the observation made by Nuth and Laloui [196] in the sense that the SIYS can be omitted from the model is valid, as it can be substituted by the maximum historical value of suction. Moreover, even if the intersection between the SIYS and the LCYS is not smooth, the volumetric strain produced by a combination of net stress and suction increase at the intersecting point is still obtained from Eq. 8.3 (Chapter 8).

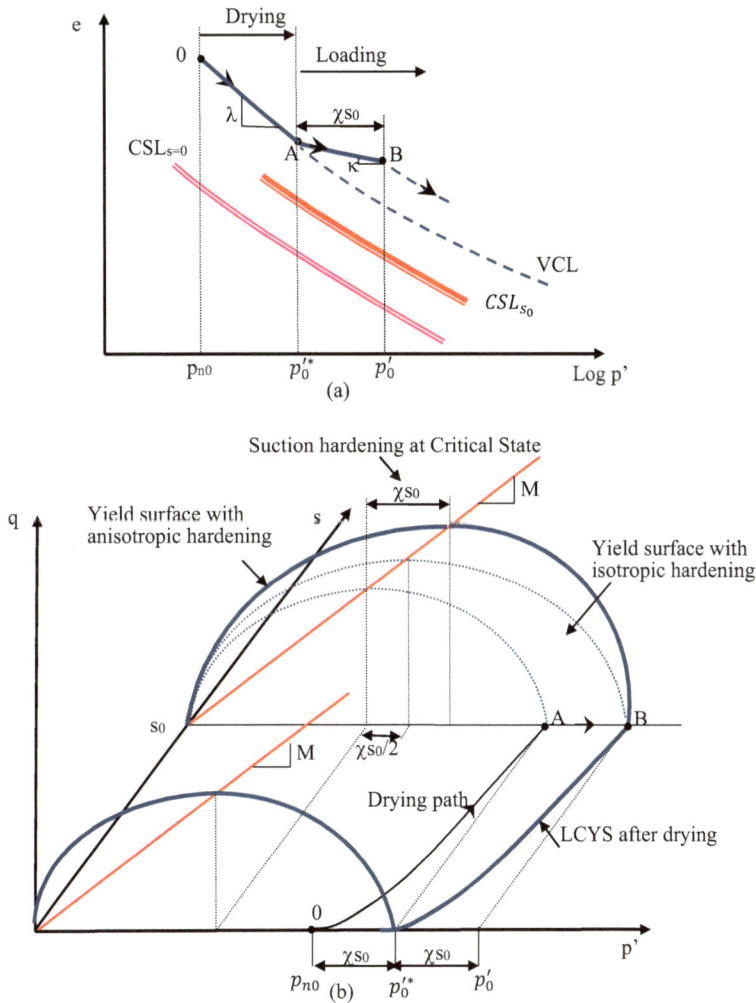

Fig. (3). (**a**) Suction hardening displaces the VCL and the CSL in the quantity χs, (**b**) the yield surface shows anisotropic hardening.

Suction hardening not only affects the position of the VCL but it pulls along the CSL as explained in the previous section. The shift of both lines is indicated in Fig. (**3a**) where $CSL_{s=0}$ and CSL_{s0} represent the position of the CSL for suction equal 0 and s_0, respectively.

Related to this phenomenon comes a second issue. When suction hardening is taken into consideration to include the shift of the VCL and CSL in the constitutive model, an anisotropic hardening of the yield surface in the effective mean stress (p') *vs.* deviator stress (q) plane occurs. Consider the elliptic yield surface of the MCCM depicted in Fig. (**3b**). When suction hardening occurs, the VCL and the CSL displaces in the quantity χs_0. However, the critical state point (center) of an ellipse showing isotropic hardening only displaces in the quantity $\chi s_0/2$. Therefore and as shown in Fig. (**3b**), in order to displace the CSL in the quantity χs_0, anisotropic hardening of the elliptic yield surface has to be considered.

A complementary phenomenon to suction hardening arises when drying is preceded by unloading in saturated conditions. Suppose that a saturated normally consolidated soil has been subjected to a maximum net stress $p_0'^*$ (point 0 in Fig. **4a**) and then unloaded to a mean net stress p_{n0} (point A in the same figure). The LCYS for the saturated overconsolidated material in the mean effective stress *vs.* suction plane can be represented by a vertical straight line located at $p_0'^*$ and indicated as $LCYS_0$ in Fig. (**4b**). Observe that in this last figure, the effective mean stress (horizontal axis) is plotted in logarithmic scale, and therefore, the drying path and the $LCYS_C$ do not run parallel to each other as shown in Fig. (**3b**). If the saturated overconsolidated sample is then subjected to drying, an initial elastic recompression is observed until suction stress reaches the preconsolidation stress $p_0'^*$ as indicated by point B in Figs. (**4a** and **b**). Only then, elastoplastic strains occur (path BC). This means that for saturated overconsolidated samples, the suction increase yield surface (SIYS) is not located at the position $s = 0$ (see Fig. **4b**) but at an equivalent suction (s_{eq}) represented by the relationship

$$s_{eq} = \frac{p_0'^* - p_{n0}}{\chi}$$

where $p_0'^*$ and p_{net} represent the saturated preconsolidation stress and the net mean stress before drying, respectively, while χ is Bishop's parameter at suction s_{eq}. This phenomenon has not been reported extensively and only few experimental results are available (see for example [93]). This phenomenon is called here saturated unloading hardening as it only occurs in saturated overconsolidated samples subjected to drying. It does not happen in unsaturated soils because water menisci restrain the elastic volumetric recovery during unloading (path BA in

Fig. **4a**). This means that when an unsaturated soil is unloaded at constant suction, it shows low volumetric recovery as some of the reduction in net stress is taken by water menisci at the contact between particles. The intake of part of the net stress by water menisci during unloading has been called the interlock or fabric stress (p_{fab}) in Chapter 9 and has been used to correctly simulate the behavior of compacted soils. The fabric stress is then the difference between the real net stress applied to the sample with the net stress required to attain the current void ratio of the sample at the corresponding suction. The fabric stress can be related to a change in the contact angle of water menisci during unloading.

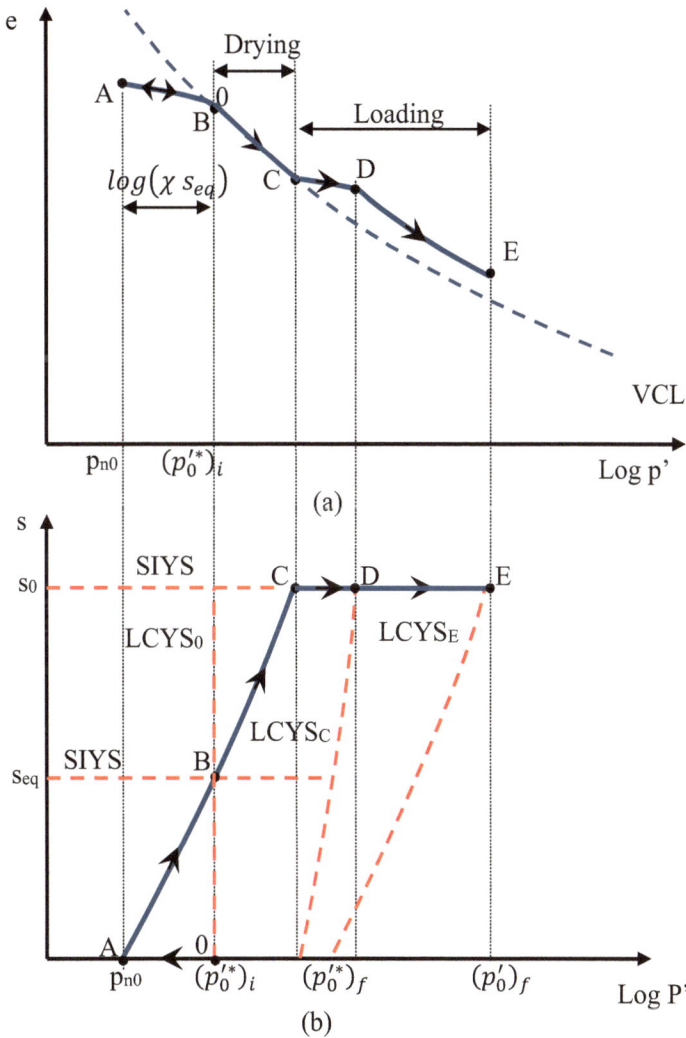

Fig. (4). Volumetric behavior of a saturated overconsolidated soil during first drying.

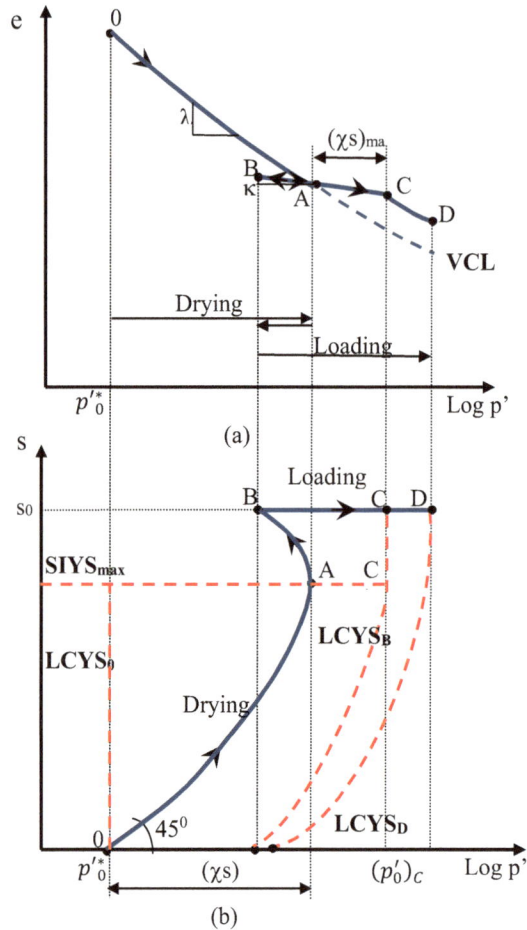

Fig. (5). Loading after elastic rebound during drying. (**a**) Volumetric behavior and (**b**) stress path and yield surfaces at different stages.

Moreover, suction stress (and therefore suction hardening) does not keep increasing with suction but it shows a maximum value and then decreases for sandy and silty soils subjected to drying ([87, 197]). Consider a saturated soil sample subjected to drying up to suction s_0. As suction increases, elastoplastic strain occurs (path 0A in Fig. **5a**) and the LCYS runs aside the drying path as indicated by line $LCYS_B$ in Fig. **5b**. If at some point, suction stress reaches its maximum value (point A, in Fig. **5**), the drying path curves to the left hand side and shows the shape sketched in Fig. (**5b**). When the maximum suction stress is reached, the apparent preconsolidation stress reaches its maximum value $(p'_0)_C$ and the SIYS is locked at this position indicated by line $SIYS_{max}$ in Fig. (**5b**).

Because further drying produces the effective stress to reduce, soil sample shows a volumetric elastic rebound (path AB in Fig. **5a**). Therefore, from the maximum suction stress and for increasing values of suction, the $LCYS_B$ shows a vertical slope (line CC' in Fig. **5b**). This means that at this stage, no further yielding occurs with increasing values of suction. This behavior has been experimentally reported by Cunningham, Ridley, Dineen and Burland [110] and Fleureau, Kheirbek-Saoud, Soemitro and Taibi [93]. If at the end of drying (point B), the soil is loaded by a net stress increase, it shows an initial elastic recompression (path BC) until the apparent preconsolidation stress is reached (point C) with a maximum suction hardening value of χs_{max}. From there on, the sample shows elastoplastic behavior (path CD). By the end of loading the LCYS has displaced to the position indicated by line $LCYS_D$.

12.4. MECHANICAL MODEL

As stated in Section 1 of this Chapter, anisotropic hardening was included in the model by Loret and Khalili [179]. This model considers a yield surface that can modify the position of the critical state with respect to the preconsolidation stress. This feature was implemented by splitting the yield surface in two ellipses that meet horizontally at the critical state point in the axes of effective mean stress *versus* deviator stress. Here a similar approach is considered except that the two curves forming the yield surface can have different shapes (not only ellipses) and therefore a large number of combinations can be considered for the geometry of the yield surface in addition to the position of the critical state. The proposed equations for the left (dilating behavior) and right side (compressive behavior) of the yield surface are

$$q^{a_1} - M^{a_1}\{(f\,p_0')^{a_1} - |p' - f\,p_0'|^{a_1}\} = 0$$

$$q^{a_2} - \left(\tfrac{M\,f}{1-f}\right)^{a_2}\left\{(p_0'(1-f))^{a_2} - |p' - f\,p_0'|^{a_2}\right\} = 0 \tag{12.1}$$

where a_1 and a_2 represent the exponent of the left and right side of the yield surface, respectively, while f is the ratio between the effective mean stress at the critical state and the preconsolidation effective stress. The left and right side of the yield surface are defined according to the position of the critical mean stress $f\,\bar{p}_0'$.

Eq. (12.1) ensures that for any combination of values where $a_1, a_2 > 1$, and $0 < f < 1$, both segments of the yield surface meet horizontally at the critical state and reach vertically the effective mean stress axis. Different combinations for parameters a_1, a_2 and f are represented in Fig. (**6**). The three numbers for each surface, represent the values of parameters a_1, a_2 and f, in this order. When few experimental results are available, the yield surface can take the usual ellipsoidal shape of the MCCM ($a_1 = a_2 = 2$) while the position of the critical state with respect to the preconsolidation stress (f) can take values in the range between 0.4 - 0.6.

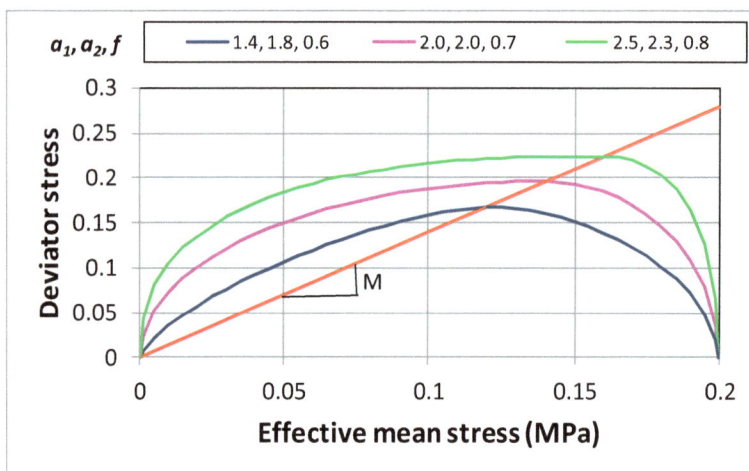

Fig. (6). Different shapes and critical state points for the yield surface in the (p', q) plane.

According to experience, the MCCM shows fair results for compressive materials (normally consolidated and slightly overconsolidated soils) but is not precise enough for dilating materials (highly overconsolidated soils), especially those concerning the volumetric strains. In that sense, the bounding surface (Dafalias and Hermann [198]) and the subloading surface concept (Hashiguchi [199]) offer an alternative to increase the precision of the model for highly overconsolidated soils as their dilating behavior can start before reaching the yield surface. In addition, the concept of sate parameter introduced by Been and Jefferies [200] for the case of sands has proved to be successful in modeling the behavior of highly overconsolidated soils. In that sense, a similar approach to that proposed by Jockovic and Vukicevic [201] is adopted here in order to increase the precision of the MCCM in modeling the behavior of highly overconsolidated soils. In that sense, an associated flow rule is considered for the model.

Notice that for the development of the model proposed herein only axisymmetric drained tests are considered. The proposed model shows two main differences with respect to other bounding surface models. The first one is that the subloading surface only appears when highly overconsolidated samples are being tested. Highly overconsolidated saturated soils tested in drained conditions, reach the left side of the elliptic yield surface following a stress path with slope 3:1 similar to the stress path 0ABA shown in Fig. (7). This drained path reaches the yield surface (point B) after crossing the CSL (point A) and finally comes back to reach the critical state at point A. The elastoplastic dilating behavior of the sample starts precisely once the stress path crosses the CSL at point A. In order to capture this behavior, a bounding and a subloading surface are considered. The positions of the bounding and subloading surfaces are given by the initial preconsolidation stress of the sample $\bar{p}'_0$, and the effective mean stress p'_0, respectively, as shown in Fig. (7). The position of the subloading surface ensures that at its critical mean stress $f\,\bar{p}'_0$, occurs the crossing of the stress path and the failure line, as indicated in Fig. (7). When the stress path reaches point A for the first time, the soil is in critical condition as the stress state lays exactly at the intersection between the subloading surface and the CSL with slope $M = 6\sin\varphi/(3 - \varphi)$. In order to avoid failure of the sample at this stage, the hardening state parameter w is introduced. This parameter also ensures that the maximum deviator stress reached at the bounding surface (point B) corresponds with the maximum dilating ratio $d_{\varepsilon v}/d_{\varepsilon q}$, where $d_{\varepsilon v}$ and $d_{\varepsilon q}$ represent the increment of the volumetric and deviatoric strains, respectively. The hardening state parameter w used herein is similar to that proposed by Jockovic and Vukicevic [201] although it is written differently, in the form

$$w = \left(\frac{\bar{\psi}}{\psi}\right)^{R} \tag{12.2}$$

where $\bar{\psi}$ and ψ represent the variation of void ratio between the CSL and the unloading-reloading line (URL) at the bounding and subloading surface, respectively, as represented in Fig. (8). $R = (p'_i/\bar{p}'_0)^2$ represents the overconsolidation ratio of the sample to the square. Notice that Fig. (8) is represented on a logarithmic plane in the axes of effective mean stress against void ratio.

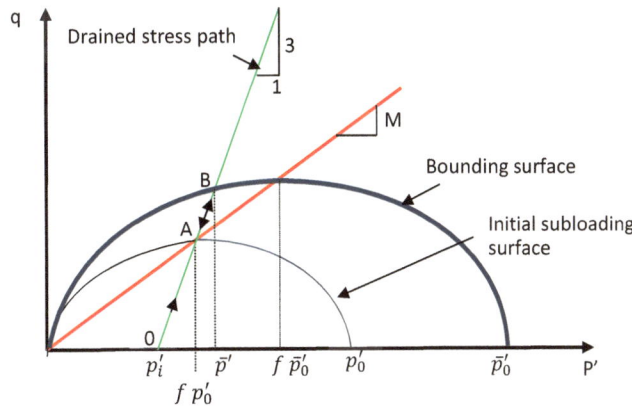

Fig. (7). Bounding surface, loading surface and stress path mapping.

The inverse of w multiplies the hardening parameter of the critical state model but only for the case of highly overconsolidated soils. At point A, parameter ψ becomes nil, w tends to infinity and, therefore, plastic strains also become nil. Then, as the stress path moves towards point B, parameter w progressively reduces until it reaches a unit value at point B. In consequence, the dilating behavior becomes maximum as well as the deviator stress at this point. When the sample loosens from point B to A, the subloading surface disappears and the state parameter becomes unit.

The second difference with respect to other models is the location of the center of the mapping rule used to establish the projection of the state of stresses on the bounding surface. While most models locate this center at the origin of the (p', q) plane, in this case, the center is located at the confining stress p_i' prior to the application of a deviator stress. Accordingly, all projections of the state of stresses on the bounding surface for highly overconsolidated soils appear on the left side of the yield surface (dilating behavior).

Under these circumstances, parameters ψ and $\bar{\psi}$ can be obtained according to the following procedure. Both the VCL and CSL are defined by the Eq. 8.4 in Chapter 8. For overconsolidated materials, exponent λ is substituted by κ which represents the slope of the URL. Therefore, parameter ψ can be obtained using the relationship

$$\psi = \log\left(\frac{e_c}{e}\right) = \log\left(\frac{f\, p_0'}{p'}\right)^{(\lambda - \kappa)} \tag{12.3}$$

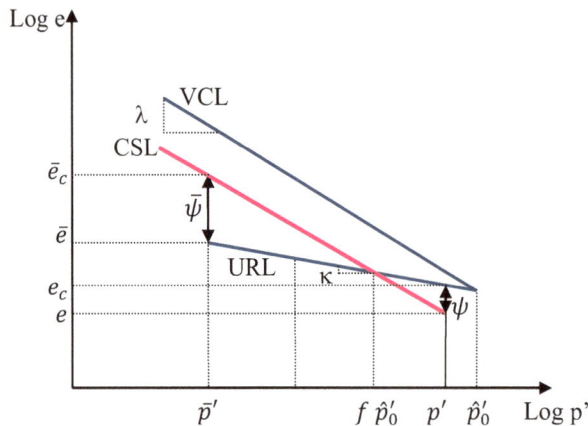

Fig. (8). Definition of parameters $\bar{\psi}$ and ψ. Parameter $\hat{p}_0'$ indicate the preconsolidation stress which can take the value of the subloading surface ($\hat{p}_0'$) or the bounding surface ($\bar{p}_0'$).

where e_c represents the void ratio at the critical state. Similarly, parameter $\bar{\psi}$ can be written as

$$\bar{\psi} = log\left(\frac{\bar{e}_c}{\bar{e}}\right) = log\left(\frac{f\,\bar{p}'_0}{\bar{p}'}\right)^{(\lambda-\kappa)} \tag{12.4}$$

where $\bar{e}$ and $\bar{e}_c$ represent the void ratio and the critical state void ratio at the projection of the current state of stresses on the bounding surface. Therefore, by substituting Eqs. (12.3) and (12.4) into (12.2) the state parameter w can be written as

$$w = \left[\frac{log\left(\frac{f\,\bar{p}'_0}{\bar{p}'}\right)}{log\left(\frac{f\,p'_0}{p'}\right)}\right]^R \tag{12.5}$$

Then, the plastic stress-strain relationship results in

$$\begin{bmatrix} d\varepsilon_v^p \\ d\varepsilon_q^p \end{bmatrix} = \frac{1}{H} \begin{bmatrix} a_x\left[M\left(\langle h_1\rangle + \langle h_2\rangle\left(\frac{f}{1-f}\right)\right)\right]^{a_x}(p'-p'_0 f)^{(a_x-1)} & a_x q^{(a_x-1)} \\ a_x q^{(a_x-1)} & \dfrac{[a_x q^{(a_x-1)}]}{a_x\left[M\left(\langle h_1\rangle + \langle h_2\rangle\left(\frac{f}{1-f}\right)\right)\right]^{a_x}(p'-p'_0 f)^{(a_x-1)}} \end{bmatrix} \begin{Bmatrix} dp' \\ dq \end{Bmatrix}$$

with

$$H = \left\{\frac{\lambda-\kappa}{v p'_0\, a_x(\langle h_1\rangle w + \langle h_2\rangle)\left[M\left(\langle h_1\rangle + \langle h_2\rangle\left(\frac{f}{1-f}\right)\right)\right]^{a_x}}\right\}$$

$$\left\{\frac{1}{[\langle h_1\rangle + \langle h_2\rangle(1-f)][p'_0(\langle h_1\rangle f + \langle h_2\rangle(1-f))]^{(a_x-1)} + (\langle h_1\rangle + \langle h_2\rangle f)(p'-p'_0 f)^{(a_x-1)}}\right\}$$

where$\langle h_1\rangle$ and $\langle h_2\rangle$ represent the step function for the left and right side of the yield surface, respectively, and can take the values 0 or 1 depending on which side the stress path hits the bounding surface. a_x represents the exponent of the corresponding side of the yield surface (left or right). The state parameter w takes the value given by Eq. (12.5) at hardening and becomes one at loosening. Notice that the stiffness matrix remains symmetric and reduces to the MCCM when $a_1 = a_2 = 2.0$ and $f = 0.5$.

The elastic behavior is considered dependent on the effective mean stress and current void ratio. In this way, the bulk (K) and shear modulus (G) are given by the equations

$$K = \frac{1+e}{e}\frac{p\prime}{\kappa} \qquad\qquad G = \frac{3(1-2v)}{2(1+v)}\frac{(1+e)}{e}\frac{p\prime}{\kappa}$$

where v represents the Poison ratio and κ the slope for the URL. Finally, the common hardening law for the yield surface is adopted here

$$dp_0\prime = \frac{vp_0\prime}{\lambda - k}d\varepsilon_v^p$$

Then, in order to simulate the behavior of unsaturated soil samples during triaxial tests, the proposed model requires the following parameters: M, λ, κ, v, a_1, a_2 and f in addition to the grain size distribution, the main wetting and drying retention curves for a certain confining stress and the initial state of the sample $(e_0, p_0\prime^*, \bar{p}_0, q_0, s_0)$.

12.5. NUMERICAL AND EXPERIMENTAL COMPARISONS

Three sets of controlled suction drained triaxial tests performed by different authors where compared with the numerical results to verify the precision of the proposed model. The first set of tests was carried out by Futai and Almeida [92]. These authors performed suction controlled triaxial tests on undisturbed tropical soil samples obtained at the depth of 1 and 5 m. The pore size distribution (PSD) of samples obtained at 1m in depth show a bi-modal structure and was chosen for the simulations presented here. The degree of saturation during sampling was of the order of 80 to 96%. Suction in the triaxial cells was controlled using the axis translation technique. Volume changes of soil samples in the triaxial cell were obtained using an automated rolling diaphragm device.

Fig. (9a) shows the experimental (Ex) SWRCs obtained by suction plate and filter paper techniques at drying (D) and wetting (W). In this same figure, the best numerical (N) fit for both curves as obtained from the porous-solid model is presented. Fig. (9b) shows the PSD adopted by the porous-solid model to reproduce the SWRCs. This same figure shows the experimental PSD of the soil obtained from scanning electron microscopy and mercury intrusion porosimetry techniques. When the PSD of the soil has been determined, it is possible to obtain the values of parameters f^s, f^u, S_w^u and χ as shown in Fig. (9c) during a drying process according to the definitions established in Chapter 2.

Some isotropic tests at different values of suction were also performed and are shown in Fig. (10). These results were used to obtain the values of parameters λ and κ. A set of saturated triaxial tests was performed on this material and reported elsewhere (Futai and Almeida [202]). These tests were used to obtain parameters M, a_1, a_2 and f of the soil. Fig. (**11a**) shows the experimental points with the numerical fitting for these tests. The adopted values of the parameters of the model are listed in Table **1**.

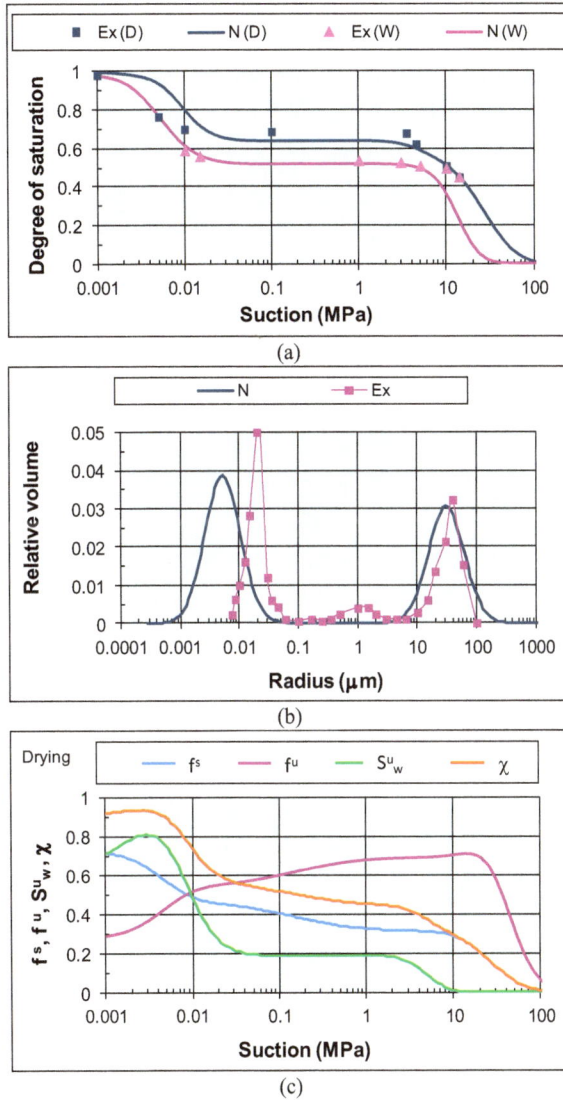

(a)

(b)

(c)

Fig. (9). (**a**) Experimental SWRCs and numerical fitting. (**b**) Experimental and numerical PSD. Experimental data from [92]. (**c**) Values of parameters f^s, f^u, f^d, S_w^u and χ.

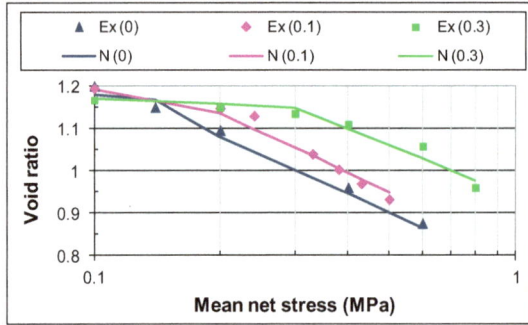

Fig. (10). Isotropic compression tests at different suctions (in MPa). Experimental data from [92].

Table 1. Parameters of the model for test reported by Futai and Almeida [202].

Parameter	M	λ	κ	v	a_1	a_2	f
Value	1.3	-0.25	-0.04	0.25	1.8	2.0	0.6

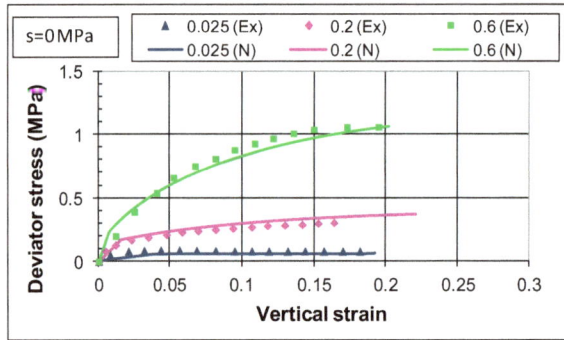

(a)

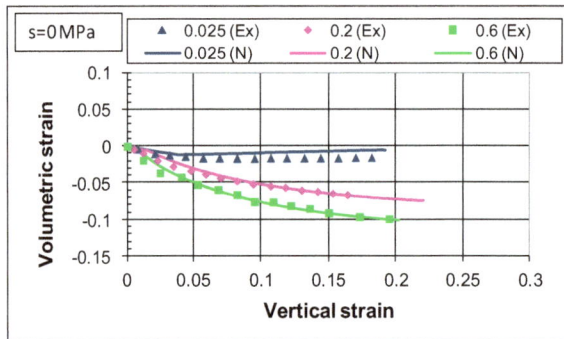

(b)

Fig. (11). Experimental results from saturated samples at different confining stresses (in MPa) and numerical fitting for model parameters. Experimental data from [202].

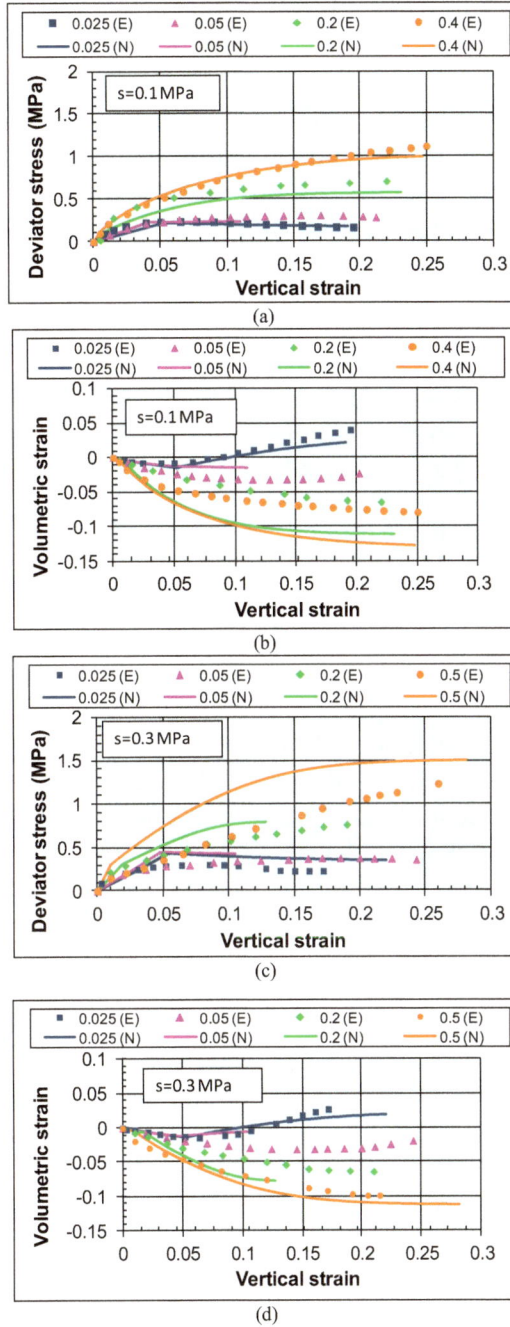

Fig. (12). Experimental and numerical results comparisons for triaxial tests at suctions of 0.1 and 0.3 MPa and different confining stresses (in MPa). Experimental data from [92].

The experimental and numerical results of tests performed at two constant values of suction (0.1 and 0.3 MPa) and different confining stresses are shown in Fig. (**12**).

Although numerical results fit well with saturated experimental data, some differences are observed when modeling unsaturated samples. In general, for samples tested at a suction of 0.1 MPa, the numerical deviator stress at failure results smaller with respect to experimental results (see Fig. **12a**) except for the test performed at a confining stress of 0.05 MPa. In contrast the predicted volumetric strain during shearing show larger values during compression (see Fig. **12b**). For samples tested at a suction of 0.3 MPa, the numerical results agree better with experimental data except for the sample tested at a confining stress of 0. 5 MPa. In that sense, the experimental results in the stress-strain plot (Fig. **10c**) show some inconsistencies related to the low stiffness and strength of this sample with respect to the other tests performed at the same suction (see Fig. **12c**).

The second set of tests was performed by Cui and Delage [203]. These authors tested an aeolian silt from Jossigny. The soil was dried, ground and sieved. Then wetted to optimum standard Proctor and leaved at rest for 24 hrs. Soil samples were compacted in three layers using a double piston system at a rate of 0.15 mm/min. Interfaces between layers were carefully scarified. The maximum compaction stress for these samples varied from 0.8 to 0.9 MPa. Initial suction of samples was measured using the filter paper techniques resulting in 0.2 MPa. Suction in triaxial tests was controlled using the osmotic technique which consisted in circulating a salty solution on top and bottom of the sample where semipermeable membranes were placed. Volume changes were measured using an inner cylinder placed around the sample and filled with colored water with a thin layer of silicon oil on top. With this arrangement volume changes were reflected on the level of oil-water interlayer which was optically monitored. Triaxial tests were performed at different suctions ($s = 0.2, 0.4, 0.8,$ and 1.5 MPa) and confining pressures ($p_n = 0.025, 0.05, 0.1, 0.2, 0.4$ and 0.6 MPa). The SWRCs for this material was reported elsewhere (Fleureau, Kheirbek-Saoud, Soemitro and Taibi [93]). The fitting of the SWRCs, as explained before, results in the determination of parameter χ as a function of suction and the volumetric strain of the sample. Some isotropic tests up to 0.6 MPa at different suctions were used to obtain the values of parameters λ and κ. Table **2** shows the adopted values of soil parameters.

Table 2. Parameters of the model for test reported by Cui and Delage [97].

Parameter	M	λ	κ	v	a_1	a_2	f
Value	1.35	-0.15	-0.025	0.25	2.0	2.0	0.5

The comparisons between experimental and numerical results are shown in Fig. (**13**) for three different confining stresses and different suctions.

From Fig. (**13**) it can be observed that numerical and experimental comparisons agree fair well for all tests except for the strength at failure for samples tested at suctions of 0.4 and 0.8 MPa and a confining stress of 0.2 MPa. The same occurs for the sample tested at a suction of 0.2 MPa and a confining stress of 0.4 MPa although the development of the volumetric strain is well simulated for these same tests.

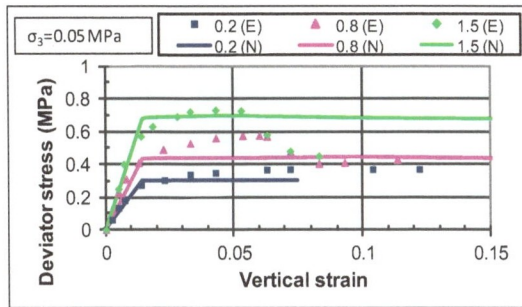

(a)

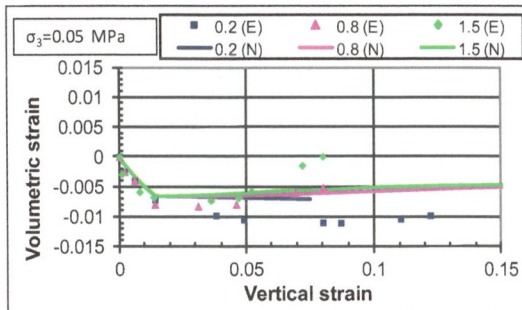

(b)

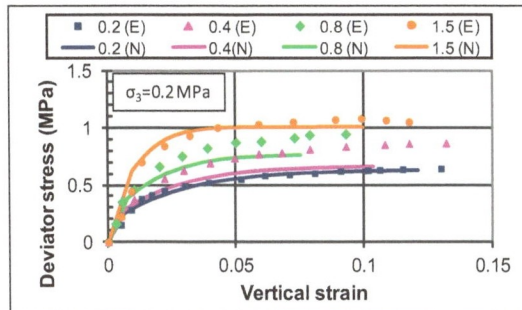

(c)

Fig. 13 cont.....

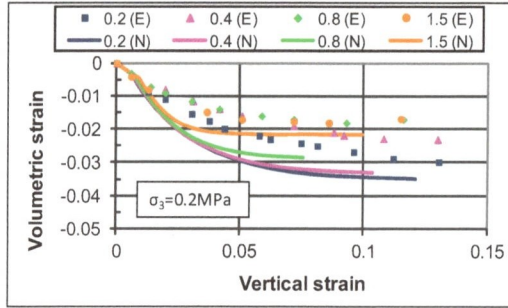

(d)

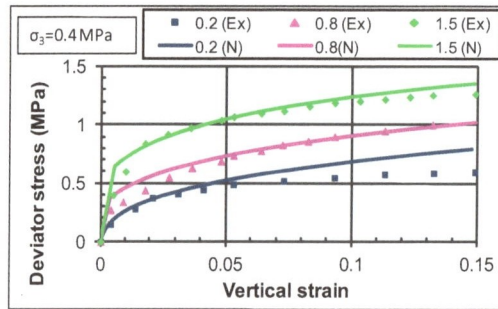

(e)

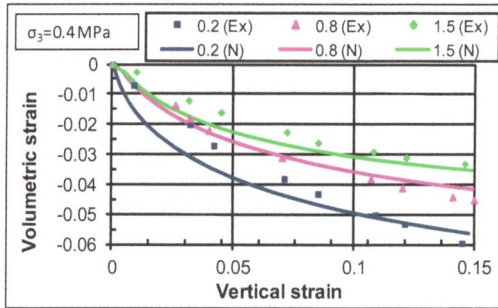

(f)

Fig. (13). Experimental and numerical results comparisons for triaxial tests performed at confining stresses of 0.05, 0.2 and 0.4 MPa and different suctions (in MPa). Experimental data from [97].

The third set of experiments was performed by Garakani, Haeri, Khosravi and Habibagahi [204]. They conducted suction controlled triaxial tests in a modified triaxial cell. Suction was controlled using the axis translation technique. Volume changes in the triaxial cell and soil sample were measured using automated volume change devices. These volume changes were corrected after a careful calibration of the equipment. Before shearing, samples were subjected to suction equilibrium followed by isotropic compression. Suctions varied from 0.1 to 0.4 MPa while isotropic stresses varied from 0.05 to 0.4 MPa. A set of saturated tests at different confining stresses were performed by the authors and are shown in Fig. (**14**) along with the numerical fitting to obtain the parameters of the model. These parameters are shown in Table **3**. The experimental and numerical results comparisons for triaxial tests performed at different suctions and confining stresses are shown in Fig. (**15**).

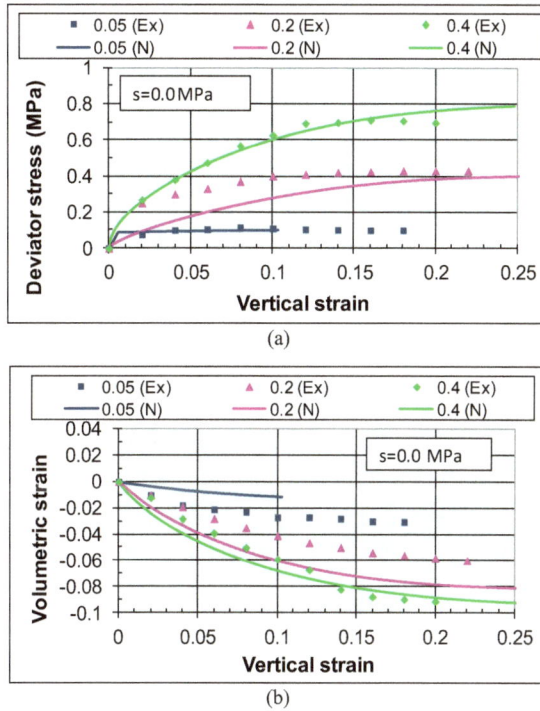

(a)

(b)

Fig. (14). Saturated tests at different confining stresses (in MPa) used to obtain the parameters of the model. Experimental data from [204].

Table 3. Parameters of the model for test reported by Garakani *et al.* [204].

Parameter	M	λ	κ	v	a_1	a_2	f
Value	1.2	-0.18	-0.02	0.25	2.0	1.8	0.5

For these tests experimental and numerical simulations show good agreement except for the strength of samples tested at a confining stress of 0.3 and suctions of 0.05 MPa and 0.4 MPa. However, the experimental results for these tests show some inconsistencies. For example, Fig. (**15c**) shows that the sample tested at a confining stress of 0.3 MPa reaches a pick strength followed by an important softening (overconsolidated behavior) whereas, the volumetric strain (Fig. **15d**) shows a continuous reduction (normally consolidated behavior) similar to that observed for sample confined at 0.15 MPa. On the contrary, the experimental

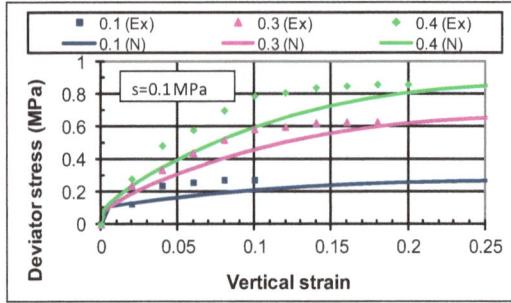

(a)

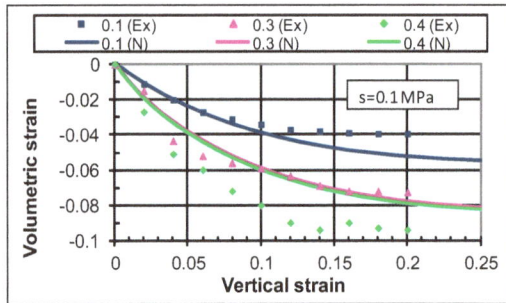

(b)

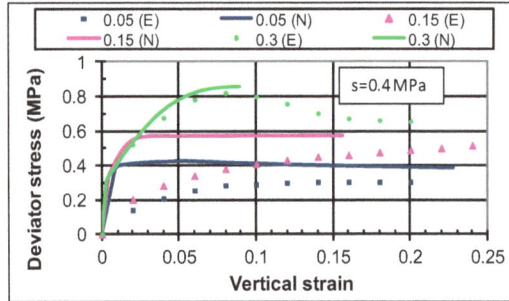

(c)

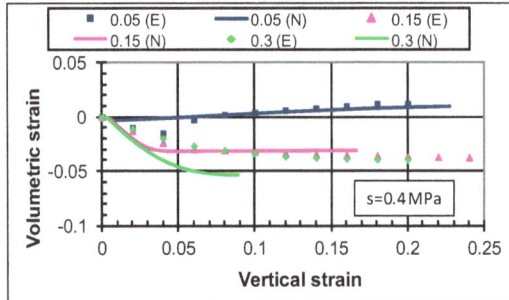

(d)

Fig. (15). Experimental and numerical results comparisons for triaxial tests performed at suctions of 0.1 and 0.4 MPa and different confining stresses (in MPa). Experimental data from [199].

volumetric strains of the sample tested at a confining stress of 0.05 MPa, shows dilation (Fig. **15d**) but no softening in the stress-strain plot (Fig. **15c**).

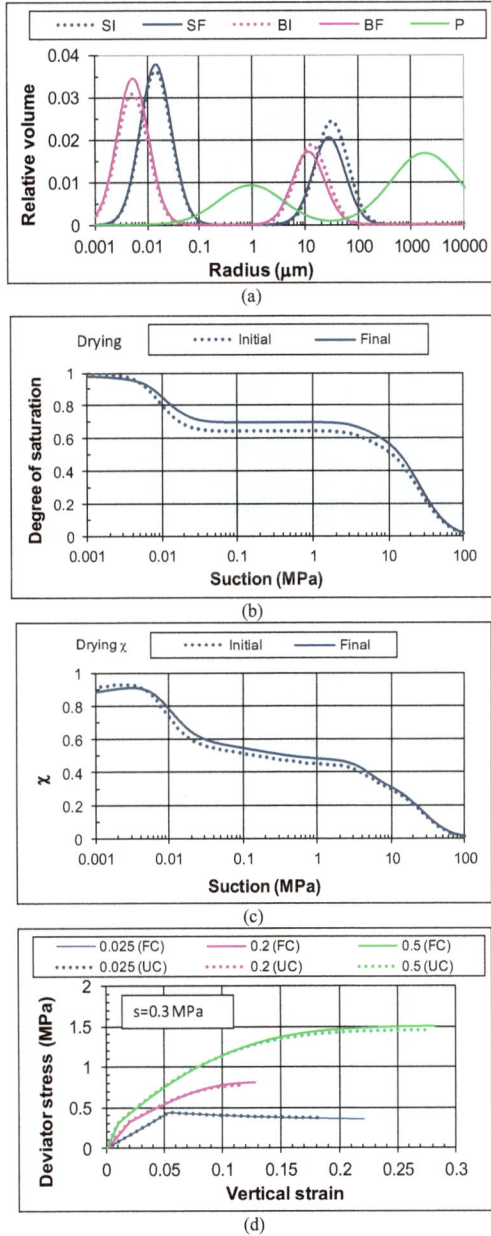

(a)

(b)

(c)

(d)

Fig. 16 cont.....

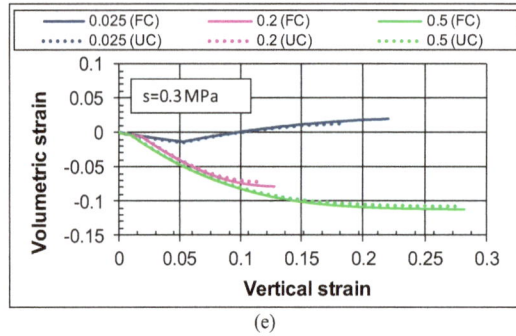

(e)

Fig. (16). (**a**) Initial and final PSD, (**b**) initial and final SWRC at drying, (**c**) initial and final values for parameter χ at drying (**d**) stress-strain behavior and (**e**) volumetric behavior for uncoupled (UC) and fully coupled (FC) model at suction a of 0.3 MPa and different confining stresses (in MPa).

In order to observe the influence of the volumetric strain on the PSD, the SWRCs and the behavior of the soil, Fig. (**16**) has been prepared. These results are based on the simulations of tests performed by Futai and Almeida [92] at a suction of 0.3 MPa and presented in Figs. (**12c** and **d**). The variation of the PSD of the sample confined at 0.5 MPa during shearing, is shown in Fig. (**16a**) and compared with the initial PSD of the sample. The initial and final distributions for sites ((SI) and (SF), respectively) and bonds ((BI) and (BF)) as well as the grain size distribution (P) of the sample are shown in this figure. It can be observed that the larger sites and bonds reduce their mean size and consequently their relative volume while the smaller elements shown an apparent increase in their relative volume as experimentally observed (Simms and Yanful [46]). Fig. (**16b**) shows the shift of the initial retention curve at the end of the test when the numerical results predict a volumetric strain of around 11% (Fig. **10d**). The influence of the volumetric strain on the values of parameter χ is shown in Fig. (**16c**). As it can be observed, values of parameter χ increase when hydro-mechanical coupling is considered. Finally, the influence on the strength and volumetric behavior of this progressive change in the values of parameter χ as the soil deforms, is shown in Figs. (**16d** and **e**), respectively. In these figures, the strength and volumetric response of the soil is represented for the fully coupled (FC) and the uncoupled (UC) model. With respect to the strength, it is observed that when hydro-mechanical coupling is taken into account both the shear stress and the volumetric strain at failure slightly increase. The increase in shear strength depends on the volumetric strain of the sample at failure. The larger the volumetric strain at failure, the larger the increase in shear strength. This is so because the shift of the SWRCs depends on the volumetric strain of the sample. This shift produces an increase of stresses during compression and a reduction during dilation. Therefore, the volumetric strain at failure increases (in absolute value) during both compression and dilation. It can be observed that the effect of hydro-mechanical coupling is rather small during monotonic loading as the retention curve slowly displaces during the deformation of the sample. Therefore, for practical applications hydro-mechanical coupling during monotonic loading can be neglected. Larger influence of hydro-mechanical coupling can be expected during large loading-unloading cycles as loading inversions initiates with the displaced retention curves on the axes of suction.

REFERENCES

[1]   E.L. Matyas, and H.S. Radhakrishna, "Volume change characteristics of partially saturated soils", *Geotechnique,* vol. 18, pp. 432-448, 1968.
[http://dx.doi.org/10.1680/geot.1968.18.4.432]

[2]   V.Q. Hung, D.G. Fredlund, and J.H.F. Pereira, "Coupled solution for the prediction of volume change in expansive soils", *Proceedings of the Third International Conference on Unsaturated Soils* Recife Brazil, vol. 1, pp. 181-186, 2002.

[3]   J.A. Jiménez, "Hacia una mecánica de suelos no saturados", Décima Conferencia Nabor Carrillo, Sociedad Mexicana de Mecánica de Suelos, México, 1990.

[4]   X. Zhang, and R.L. Lytton, "Modified state-surface approach to the study of unsaturated soil behavior. Part I: Basic concept", *Can. Geotech. J.,* vol. 46, pp. 536-552, 2009.
[http://dx.doi.org/10.1139/T08-136]

[5]   D.G. Fredlund, and N.R. Morgenstern, "Stress state variables for unsaturated soils", *J. Geotech. Eng. Div.,* vol. 103, pp. 447-466, 1977.

[6]   A.W. Bishop, and I.B. Donald, "The experimental study of partly saturated soil in the triaxial apparatus", *Proc. 5th Int. Conf. Soil Mech.,* vol. I, 1961pp. 13-21 Paris.

[7]   E.E. Alonso, A. Gens, and A. Josa, "A constitutive model for partially saturated soils", *Geotechnique,* vol. 40, pp. 405-430, 1990.
[http://dx.doi.org/10.1680/geot.1990.40.3.405]

[8]   K.H. Roscoe, and J.B. Burland, "On the generalized stress-strain behaviour of wet clay", Engineering Plasticity, De Heyman and Leckie, Cambridge University Press, pp: 535-609, 1968.

[9]   K. Terzaghi, "The shearing resistance of saturated soils and the angle between the planes of shear", in Proc. 1st Int. Conf. Soil Mech., Int. Soc. Soil Mech. Found. Engrng, vol. 1, pp. 54-56, 1936.

[10]   A.W. Skempton, Terzaghi's discovery of effective stress. *From theory to practice in soil mechanics.* John Wiley: New York, USA, 1960.

[11]   P.V. Lade, and R. De Boer, "The concept of effective stress for soil, concrete and rock", *Geotechnique,* vol. 47, pp. 61-78, 1997.
[http://dx.doi.org/10.1680/geot.1997.47.1.61]

[12]   J.E.B. Jennings, "Discussion on M.S. Youssef's paper", *Proc. 4th Int. Conf. on Soil Mech., Int. Soc. Soil Mech. Found. Engrng.,* vol. 3, 1957p. 168

[13]   D. Croney, J.D. Coleman, and W.P.M. Black, "Movement and distribution of water in soil in relation to highway design and performance", Highway Research Board, Spec. Report No. 40, 1958.

[14]   A.W. Bishop, "The principle of effective stress", *Tek. Ukebl.,* vol. 39, pp. 859-863, 1959.

[15]   G.D. Aitchison, "Relationships of moisture stress functions in unsaturated soils", In: *Conf. Pore Pressures, Institution of Civil Engineering.* Buttherworths: London, U.K, 1960.

[16]   G.E. Blight, "Effective stress evaluation for unsaturated soils", *J. Soil Mech. Div,* vol. 93, pp. 125-148, 1967.

[17]   N. Khalili, and M.H. Khabbaz, "A unique relationship for χ for the determination of the shear strength of unsaturated soils", *Geotechnique,* vol. 48, pp. 681-687, 1998.
[http://dx.doi.org/10.1680/geot. 1998.48.5.681]

[18]   A-L. Öberg, and G. Sällfors, "A rational approach to the determination of the shear strength parameters of unsaturated soils", *Proceedings 1ˢᵗ Int. Conf. on Unsaturated Soils, Alonso & Delage eds*, vol. 1, 1995pp. 151-158 Paris, France.

[19]   E.A. Garven, and S.K. Vanapalli, "Evaluation of empirical procedures for predicting the shear strength of unsaturated soils", *Proceedings 5ᵗʰ International Congress on Unsaturated Soil Mechanics*, 2006 Arizona, USA
       [http://dx.doi.org/10.1061/40802(189)219]

[20]   J.E.B. Jennings, and J.B. Burland, "Limitations to the use of effective stress in partly saturated soils", *Geotechnique*, vol. 12, pp. 125-144, 1962.
       [http://dx.doi.org/10.1680/geot.1962.12.2.125]

[21]   S.K. Vanapalli, D.G. Fredlund, D.E. Pufahl, and A.W. Clifton, "Model for the prediction of shear strength with respect to soil suction", *Can. Geotech. J.*, vol. 33, pp. 379-392, 1996.
       [http://dx.doi.org/ 10.1139/t96-060]

[22]   S.J. Wheeler, R.S. Sharma, and M.S.R. Buisson, "Coupling hydraulic hysteresis and stress-strain behaviour in unsaturated soils", *Geotechnique*, vol. 53, pp. 41-54, 2003.
       [http://dx.doi.org/ 10.1680/geot.2003.53.1.41]

[23]   D. Gallipoli, A. Gens, and R. Sharma, *Geotechnique*, vol. 53, pp. 123-135, 2003.
       [http://dx.doi.org/ 10.1680/geot.2003.53.1.123]

[24]   R. Tamagnini, "An extended Cam-clay model for unsaturated soils with hydraulic hysteresis", *Geotechnique*, vol. 54, pp. 223-228, 2004.
       [http://dx.doi.org/10.1680/geot.2004.54.3.223]

[25]   D. Sheng, D.G. Sloan, and A. Gens, "A constitutive model for unsaturated soils: thermomechanical and computational aspects", *Comput. Mech.*, vol. 33, pp. 453-465, 2004.
       [http://dx.doi.org/10.1007/ s00466-003-0545-x]

[26]   C.S. Desai, and Z. Wang, "Disturbed state model for porous saturated materials", *Int. J. Geomech.*, vol. 3, pp. 260-265, 2003.
       [http://dx.doi.org/10.1061/(ASCE)1532-3641(2003)3:2(260)]

[27]   A. Sridharan, A.G. Altschaeffl, and S. Diamond, "Pore size distributions studies", *J. Soil Mech. Found. Div.*, vol. 97, pp. 771-787, 1971.

[28]   W.B. Haines, "The hysteresis effect in capillary properties and the mode of moisture distribution associated therewith", *J. Agric. Sci.*, vol. 20, p. 7, 1929.

[29]   E.J. Murray, "An equation of state for unsaturated soils", *Can. J. Geotech*, vol. 39, pp. 125-140, 2002.
       [http://dx.doi.org/10.1139/t01-087]

[30]   C.S. Desai, and Z. Wang, "Disturbed state model for porous saturated materials", *Int. J. Geomech.*, vol. 3, pp. 260-265, 2003.
       [http://dx.doi.org/10.1061/(ASCE)1532-3641(2003)3:2(260)]

[31]   A-L. Öberg, Stability of sand and silt slopes.*Internal report.* Department of Geotechnical Engineering, Chalmers University of Technology: Gothenburg, Sweden, 1995.

[32]   M.A. Biot, "Theory of elasticity and consolidation for a porous anisotropic solid", *J. Appl. Phys.*, vol. 26, pp. 182-185, 1955.
       [http://dx.doi.org/10.1063/1.1721956]

[33]   W.B. Haines, "A note on the cohesion developed by capillary forces in an ideal soil", *J. Agric. Sci.*, vol. 15, pp. 529-535, 1925.
       [http://dx.doi.org/10.1017/S0021859600082460]

[34]   V. Escario, J.F.T. Jucá, and M.S. Coppe, "Strength and deformation of partly saturated soils", *Proceedings of the 12ᵗʰ International Conference of Soil Mechanics and Foundation Engineering*, vol. 1, 1989pp. 43-49 Río de Janeiro, Brazil

[35]   J.K. Gan, and D.G. Fredlund, "Shear strength characteristics of two saprolitic soils", *Can. Geotech. J.,* vol. 25, pp. 500-510, 1996.
[http://dx.doi.org/10.1139/t88-055]

[36]   A.W. Bishop, and G.E. Blight, "Some aspects of effective stress in saturated and partly saturated soils", *Geotechnique,* vol. 13, pp. 177-197, 1963.
[http://dx.doi.org/10.1680/geot.1963.13.3.177]

[37]   M.M. Allan, and A. Sridharan, "Effect of wetting and drying on shear strength", *J. Geotech. Eng. Div.,* vol. 107, pp. 421-438, 1981.

[38]   T. Nishimura, Y. Hirabayashi, D.G. Fredlund, and J. Gan, "Influence of stress history on the strength parameters of an unsaturated statically compacted soil", *Can. Geotech. J.,* vol. 36, pp. 251-261, 1999.
[http://dx.doi.org/10.1139/t98-098]

[39]   V. Sivakumar, and S.J. Wheeler, "Influence of compaction procedure on the mechanical behaviour of an unsaturated compacted clay (Part 1 and 2)", *Geotechnique,* vol. 50, pp. 359-376, 2000.
[http://dx.doi.org/ 10.1680/geot.2000.50.4.359]

[40]   G. Klubertanz, L. Laloui, L. Vulliet, and P. Gachet, "Experimental validation of the hydromechanical modeling of unsaturated soils", *Proceedings Workshop Chemo-mechanical coupling in clays,* 2002pp. 223-230

[41]   V. Mayagoitia, F. Rojas, and I. Kornhauser, *J. Chem. Soc., Faraday Trans.,* vol. 84, p. 785, 1988.
[http://dx.doi.org/10.1039/f19888400785]

[42]   V. Mayagoitia, and I. Kornhauser, Fundamentals of adsorption IV, Suzuki M. ed., The engineering foundation, New York, USA, 1990.

[43]   D.G. Fredlund, and A. Xing, "Equations for the soil-water characteristic curve", *Can. Geotech. J.,* vol. 31, pp. 521-532, 1994.
[http://dx.doi.org/10.1139/t94-061]

[44]   P.H. Simms, and E.K. Yanful, "Pore network modelling for unsaturated soils", *Geotech. Conf.,* 2003 Winnipeg, Canada.

[45]   P.H. Simms, and E.K. Yanful, "Measurement and estimation of pore shrinkage and pore distribution in a clayey till during soil-water characteristic curve tests", *Can. Geotech. J.,* vol. 38, pp. 741-754, 2001.
[http://dx.doi.org/10.1139/t01-014]

[46]   P.H. Simms, and E.K. Yanful, "A discussion of the application of mercury intrusion porosimetry for the investigation of soils, including an evaluation of its use to estimate volume change in compacted clays", *Geotechnique,* vol. 54, pp. 421-426, 2004.
[http://dx.doi.org/10.1680/geot.2004.54.6.421]

[47]   N.R. Morrow, "Physics and thermodynamics of capillary action in Porous media", *Ind. Eng. Chem.,* vol. 62, p. 32, 1970.
[http://dx.doi.org/10.1021/ie50726a006]

[48]   D.H. Everet, ""The solid–gas interface", E", *Edisson Flood, Dekker, New York, USA,* vol. II, pp. 1005-1010, 1967.

[49]   R.P. Ray, and K.B. Morris, "Automated laboratory testing for soils-water characteristic curves", *Proc. 1ˢᵗ Int. Conf. Unsat.,* vol. 1, 1995pp. 547-552 Paris, France.

[50]   P.A. Cundall, and O.D.L. Strack, "The development of constitutive laws for soil using the Distinct Element Method", *Proceedings Int. Conf. Num. Meth. Geomech,* vol. 1, pp. 289-298, 1979.

[51]   J.A. Gili, and E.E. Alonso, "Microstructural deformation mechanisms of unsaturated granular soils", *Int. J. Numer. Anal. Methods Geomech.,* vol. 26, pp. 1-36, 2002.
[http://dx.doi.org/10.1002/nag.206]

[52]   E.E. Alonso, E. Rojas, and N.M. Pinyol, *"Unsaturated soil mechanics", Reunión Nacional de*

Mecánica de Suelos. vol. 117-205. Especial: Aguascalientes, Mexico, 2008.

[53] G.A. Leonards, A. Alarcon, J.D. Frost, Y.E. Mohamedzein, J.C. Santamarina, S. Thevanayagam, J.E. Tomaz, and J.E. Tyree, "Dynamic penetration resistance and the prediction of the compressibility of a fine-grained sand-A laboratory study", *Discussion, Géotechnique,* vol. 36, pp. 275-279, 1986.
[http://dx.doi.org/10.1680/geot.1986.36.2.275]

[54] J. Horta, E. Rojas, M. L. Pérez, T. López-Lara, and J. B. Hernández, "A random porous model to simulate the retention curve of soils", *Int. J. Num. Anal. Meth. Geomech..* 2012.

[55] F.A.L. Dullien, *Porous media, fluid transport and pore structure.* Academic Press: USA. 1992.

[56] D.W. Taylor, Fundamentals of soil mechanics, John Wiley ed., New York, USA, 1954.

[57] K. Collins, and A. McGown, "The form and function of microfabric features in a variety of natural soils", *Geotechnique,* vol. 24, pp. 223-254, 1974.
[http://dx.doi.org/10.1680/geot.1974.24.2.223]

[58] M. He, A. Szuchmacher, D.E. Aston, C. Buenviaje, R. Overney, and R. Luginbuhl, "Critical phenomenon of water bridges in nanoasperity contacts", *J. Chem. Phys.,* vol. 114, pp. 1355-1360, 2001.
[http://dx.doi.org/10.1063/1.1331298]

[59] E. Rojas, and F. Rojas, "Modeling hysteresis of the soil-water characteristic curve", *Soil Found.,* vol. 45, pp. 135-146, 2005.
[http://dx.doi.org/10.3208/sandf.45.3_135]

[60] W.B. Haines, "Further contribution to the theory of capillary phenomena in soil", *J. Agric. Sci.,* vol. 17, pp. 264-290, 1927.
[http://dx.doi.org/10.1017/S0021859600018499]

[61] M.T. van Genuchten, "A closed form equation for predicting the hydraulic conductivity of unsaturated soils", *Soil Sci. Soc. Am. J.,* vol. 44, pp. 892-898, 1980.
[http://dx.doi.org/10.2136/sssaj1980.0361599500 4400050002x]

[62] A.J. Brown, "The thermodynamics and hysteresis of adsorption", PhD thesis, University of Bristol, England, 1963.

[63] B.V. Enustun, and M. Enuysal, "Determination of the Pore Size Distribution by Direct Methods", *Middle East Tech. Univ. J. Pure Applied Sci.,* vol. 3, pp. 81-88, 1970.

[64] G. Mason, *Determination of the Pore Size Distribution and Pore-Space Interconnectivity of Vycor Porous Glass from Adsorption-Desorption Hysteresis Capillary Condensation Isotherms*, 1988.

[65] D. Penumadu, and J. Dean, "Compressibility effect in evaluating the pore-size distribution of kaolin clay using mercury intrusion porosimetry", *Can. Geotech. J.,* vol. 37, pp. 393-405, 2000.
[http://dx.doi.org/ 10.1139/t99-121]

[66] E. Romero, A. Gens, and A. Lloret, "Water permeability, water retention and microstructure of unsaturated compacted Boom clay", *Eng. Geol.,* vol. 54, pp. 117-127, 1999.
[http://dx.doi.org/ 10.1016/S0013-7952(99)00067-8]

[67] A.B. Abell, K.L. Willis, and D.A. Lange, "Mercury intrusion porosimetry and image analysis of cement-based materials", *J. Colloid Interface Sci.,* vol. 211, no. 1, pp. 39-44, 1999.
[http://dx.doi.org/ 10.1006/jcis.1998.5986] [PMID: 9929433]

[68] S. Roels, J. Elsen, J. Cermeliet, and H. Hens, "Characterization of pore structure by combining mercury porosimetry and micrography", *Mater. Struct.,* vol. 34, pp. 76-82, 2001.
[http://dx.doi.org/ 10.1007/BF02481555]

[69] A. Sawangsuriya, T.B. Edil, and P.J. Bosscher, "Modulus-suction-moisture relationship for compacted soils in postcompaction state", *J. Geotech. Geoenviron. Eng.,* vol. 135, pp. 1390-1403, 2009.
[http://dx.doi.org/ 10.1061/(ASCE)GT.1943-5606.0000108]

[70] L.F. Vesga, "Equivalent effective stress and compressibility of unsaturated kaolinite clay subjected to drying", *J. Geotech. Geoenviron. Eng.,* vol. 134, pp. 366-378, 2008.
[http://dx.doi.org/10.1061/ (ASCE)1090-0241(2008)134:3(366)]

[71] D.G. Fredlund, A. Xing, M.D. Fredlund, and S.L. Barbour, "The relationship of unsaturated soil shear strength to the soil–water characteristic curve", *Can. Geotech. J.,* vol. 33, pp. 440-448, 1996.
[http://dx.doi.org/10.1139/t96-065]

[72] M.D. Fredlund, D.G. Fredlund, and G. Wilson, "Prediction of the soil-water characteristic curves from grain-size distribution and volume-mass properties", *Proc. Third Brazilian Symp. Unsat. Soils,* vol. 1, 1997pp. 13-23 Rio de Janeiro, Brazil

[73] M. Aubertin, J.F. Richard, and R.P. Chapuis, "A predictive model for the water retention curve: Applications to tailings from hard rock mines", *Can. Geotech. J.,* vol. 35, pp. 55-69, 1998.
[http://dx.doi.org/ 10.1139/t97-080]

[74] J.S.C. Mbagwu, and C.N. Mbah, "Estimating water retention and availability in Nigerian soils from their saturation percentage", *Commun. Soil Sci. Plant Anal.,* vol. 29, pp. 913-922, 1998.
[http://dx.doi.org/ 10.1080/00103629809369995]

[75] L.M. Arya, and J.F. Paris, "Physicoempirical model to predict the soil moisture characteristic from particle-size distribution and bulk density", *Soil Sci. Soc. Am. J.,* vol. 45, pp. 1023-1030, 1981.
[http://dx.doi.org/10.2136/sssaj1981.03615995004500060004x]

[76] L.M. Arya, and T.S. Dierolf, "Predicting soil moisture characteristics from particle size distributions: an improved method to calculate pore radii from particle radii", *Proc. Int. Workshop Indirect Meth. Estim. Hydraulic Prop. Unsat. Soils,* 1992pp. 115-124

[77] A. Basile, and G. D'Urso, "Experimental corrections of simplified methods for predicting water retention curves in clay-loamy soils from particle size determination", *Soil Technol.,* vol. 10, pp. 261-272, 1997.
[http://dx.doi.org/10.1016/S0933-3630(96)00020-7]

[78] P.H. Simms, and E.K. Yanful, "Predicting soil-water characteristic curves of compacted plastic soils from measured pore-size distributions", *Geotechnique,* vol. 52, pp. 269-278, 2002.
[http://dx.doi.org/ 10.1680/geot.2002.52.4.269]

[79] P.H. Simms, and E.K. Yanful, "A pore network model for hydromechanical coupling in unsaturated compacted clayey soils", *Can. Geotech. J.,* vol. 42, pp. 499-514, 2005.
[http://dx.doi.org/10.1139/t05-002]

[80] G.P. Androutsopoulos, and R. Mann, "Evaluation of mercury porosimeter experiments using a network pore structure model", *Chem. Eng. Sci.,* vol. 34, pp. 1203-1212, 1979.
[http://dx.doi.org/ 10.1016/0009-2509(79)85151-9]

[81] L.M. Zhang, and X. Li, "Microporosity structure of coarse granular soils", *J. Geotech. Geoenviron. Eng.,* vol. 136, pp. 1425-1436, 2010.
[http://dx.doi.org/10.1061/(ASCE)GT.1943-5606.0000348]

[82] J. Bear, "Hydraulic of Groundwater." McGraw-Hill, Series of Water Resources and Environmental Eng., 1979.

[83] S. Wheeler, and V. Sivakumar, "An elasto-plastic critical state framework for unsaturated soils", *Geotechnique,* vol. 45, pp. 35-53, 1995.
[http://dx.doi.org/10.1680/geot.1995.45.1.35]

[84] R. Thom, R. Sivakumar, V. Sivakumar, E.J. Murray, and P. Mackinnon, "Pore size distribution of unsaturated compacted kaolin: the initial states and states following saturation", *Geotechnique,* vol. 57, pp. 469-474, 2007.
[http://dx.doi.org/10.1680/geot.2007.57.5.469]

[85] J. Espitia, "Micromechanical model to reproduce the soil-water retention curve of soils", Master

thesis, University of Queretaro, Mexico, 2005.

[86]   V. Sivakumar, "A critical state framework for unsaturated soils", PhD thesis, University of Sheffield, UK, 1993.

[87]   E. Rojas, "An Equivalent stress equation for unsaturated soils, II: Solid-porous model", *Int. J. Geomech.,* vol. 8, pp. 291-299, 2008.
[http://dx.doi.org/10.1061/(ASCE)1532-3641(2008)8:5(291)]

[88]   A. Koliji, L. Laloui, O. Cusinier, and L. Vulliet, "Suction induced effects on the fabric of a structured soil", *Transp. Porous Media,* vol. 64, pp. 261-278, 2006.
[http://dx.doi.org/10.1007/s11242-005-3656-3]

[89]   L.F. Vesga, and L.E. Vallejo, "Direct and indirect tensile tests for measuring the equivalent effective stress in a kaolinite clay", *Proceedings Fourth Int. Conf. Unsat. Soils,* vol. 1, 2006pp. 1290-13 Arizona, USA
[http://dx.doi.org/10.1061/40802(189)106]

[90]   J. Vaunat, E. Romero, and C. Jommi, *An elastoplastic hydromechanical model for unsaturated soils,* .

[91]   D.A. Sun, H.B. Cui, H. Matsuoka, and D. Sheng, "A three-dimensional elastoplastic model for unsaturated compacted soils with hydraulic hysteresis", *Soil Found.,* vol. 47, pp. 253-264, 2007.
[http://dx.doi.org/ 10.3208/sandf.47.253]

[92]   M.M. Futai, and S.S. Almeida, "An experimental investigation of the mechanical behavior of an unsaturated gneiss residual soil", *Geotechnique,* vol. 55, pp. 201-213, 2005.
[http://dx.doi.org/ 10.1680/geot.2005.55.3.201]

[93]   J.M. Fleureau, S. Kheirbek-Saoud, R. Soemitro, and S. Taibi, "Behavior of clayey soils on drying-wetting paths", *Can. Geotech. J.,* vol. 30, pp. 287-296, 1993.
[http://dx.doi.org/10.1139/t93-024]

[94]   D. Sheng, "Review of fundamental principles in modeling unsaturated soil behavior", *Comput. Geotech.,* vol. 38, pp. 757-776, 2011.
[http://dx.doi.org/10.1016/j.compgeo.2011.05.002]

[95]   T.M. Thu, H. Rahardjo, and E-C. Leong, "Elastoplastic model for unsaturated soil with incorporation of the soil-water characteristic curve", *Can. Geotech. J.,* vol. 44, pp. 67-77, 2007.
[http://dx.doi.org/ 10.1139/t06-091]

[96]   Y. Khogo, M. Nikano, and T. Miyazaky, "Theoretical aspects of constitutive modeling for unsaturated soils", *Soil Found.,* vol. 33, pp. 49-63, 1993.
[http://dx.doi.org/10.3208/sandf1972.33.4_49]

[97]   B. Loret, and N. Khalili, "A three phase model for unsaturated soils", *Int. J. Numer. Anal. Methods Geomech.,* vol. 24, pp. 893-927, 2000.
[http://dx.doi.org/10.1002/1096-9853(200009) 24:11<893::AID-NAG105>3.0.CO;2-V]

[98]   R. Kholer, and G. Hofstetter, "A cap model for partially saturated soils", *Int. J. Numer. Anal. Methods Geomech.,* vol. 32, pp. 981-1004, 2008.
[http://dx.doi.org/10.1002/nag.658]

[99]   A. Koliji, L. Laloui, and L. Vulliet, "Constitutive modeling of unsaturated aggregated soils", *Int. J. Numer. Anal. Methods Geomech.,* vol. 34, pp. 1846-1876, 2010.
[http://dx.doi.org/10.1002/nag.888]

[100]   D. Sheng, D.G. Fredlund, and A. Gens, "A new modelling approach for unsaturated soils using independent stress variables", *Can. Geotech. J.,* vol. 45, pp. 511-534, 2008.
[http://dx.doi.org/ 10.1139/T07-112]

[101]   D. Sun, D. Sheng, and Y. Xu, "Collapse behavior of unsaturated compacted soil with different initial densities", *Can. Geotech. J.,* vol. 44, pp. 673-686, 2007.
[http://dx.doi.org/10.1139/t07-023]

[102] R.I. Borja, "Cam clay plasticity, Part V: "A mathematical framework for three phase deformation and strain localization analysis of partially saturated porous media", *Comput. Methods Appl. Mech. Eng.,* vol. 193, pp. 5301-5338, 2002.
[http://dx.doi.org/10.1016/j.cma.2003.12.067]

[103] L.R. Hoyos, and P. Arduino, "Implicit algorithms in modeling unsaturated soil response in three-invariant stress space", *Int. J. Geomech.,* vol. 8, pp. 266-273, 2008.
[http://dx.doi.org/ 10.1061/(ASCE)1532-3641(2008)8:4(266)]

[104] H.W. Zhang, and L. Zhou, "Implicit integration of chemo-plastic constitutive model for partially saturated soils", *Int. J. Numer. Anal. Methods Geomech.,* vol. 32, pp. 1715-1735, 2008.
[http://dx.doi.org/ 10.1002/nag.690]

[105] E. Juárez-Badillo, "Constitutive relationships for soils", *Proceedings of the Symposium on Recent Developments in the Analysis of Soil Behavior and their Application to Geotechnical Structures,* 1975pp. 231-257

[106] R. Butterfield, "A natural compression law for soils (an advance on e-log p')", *Geotechnique,* vol. 29, pp. 469-480, 1979.
[http://dx.doi.org/10.1680/geot.1979.29.4.469]

[107] D. Sheng, Y. Yao, and J.P. Carter, "A volume-stress model for sands under isotropic and critical stress states", *Can. Geotech. J.,* vol. 45, pp. 1639-1645, 2008.
[http://dx.doi.org/10.1139/T08-085]

[108] J.A. Infante Sedano, and S.K. Vanapalli, "The relationship between the critical state shear strength of unsaturated soils and the soil-water characteristic curve", *Proceedings of the Fifth International Conference on Unsaturated Soils,* vol. 1, 2010pp. 253-258 Barcelona, Spain
[http://dx.doi.org/ 10.1201/b10526-31]

[109] E. Romero, A. Gens, and A. Lloret, "Suction effects on a compacted clay under non-isothermal conditions", *Geotechnique,* vol. 53, pp. 65-81, 2003.
[http://dx.doi.org/10.1680/geot.2003.53.1.65]

[110] M.R. Cunningham, A.M. Ridley, K. Dineen, and J.B. Burland, "The mechanical behaviour of a reconstituted unsaturated silty clay", *Geotechnique,* vol. 53, pp. 183-194, 2003.
[http://dx.doi.org/ 10.1680/geot.2003.53.2.183]

[111] N. Khalili, F. Geiser, and G.E. Bligth, "Effective stress in unsaturated soils: review with new evidence", *Int. J. Geomech.,* vol. 4, pp. 115-126, 2004.
[http://dx.doi.org/10.1061/(ASCE) 1532-3641(2004)4:2(115)]

[112] I. Vlahinic, H.M. Jennings, and J.J. Thomas, "A constitutive model for drying of partially saturated porous materials", *Mech. Mater.,* vol. 41, pp. 319-328, 2009.
[http://dx.doi.org/10.1016/ j.mechmat.2008.10.011]

[113] G.E. Blight, "Shrinkage during wetting of fined-pored materials. Does this accord with the principle of effective stress?", *Proceedings of the fifth International Conference on Unsaturated Soils,* vol. 1, 2010pp. 205-209
[http://dx.doi.org/10.1201/b10526-23]

[114] A. Pereira, C. Feuerharmel, W.Y.Y. Gheling, and A.V.D. Bica, "A study on the shear strength envelope of an unsaturated colluviums soil", *Proceedings of the Fourth International Conference on Unsaturated Soils,* vol. 1, 2006pp. 1191-1199 Arizona, USA
[http://dx.doi.org/10.1061/40802(189)97]

[115] D. Sheng, A. Zhou, and D.G. Fredlund, "Shear strength criteria for unsaturated soils", *Geotech. Geol. Eng.,* vol. 29, pp. 145-159, 2009.
[http://dx.doi.org/10.1007/s10706-009-9276-x]

[116] D.G. Toll, and B.H. Ong, "Critical state parameters for an unsaturated residual sandy clay", *Geotechnique,* vol. 53, pp. 93-103, 2005.

[http://dx.doi.org/10.1680/geot.2003.53.1.93]

[117] J.C.B. Benatti, M.G. Miguel, R.A. Rodriguez, and O.M. Vilar, "Collapsibility study for tropical soil profile using oedometric tests with controlled suction", *International Conference on Unsaturated Soils*, vol. 1, 2010pp. 193-198 Barcelona, Spain
[http://dx.doi.org/10.1201/b10526-21]

[118] "Elastoplastic soil constitutive laws generalized to partially saturated states", *Geotechnique*, vol. 46, pp. 279-289, 1996.
[http://dx.doi.org/10.1680/geot.1996.46.2.279]

[119] D. Karube, and K. Kawai, "The role of pore water in the mechanical behavior of unsaturated soils", *Geotech. Geol. Eng.*, vol. 19, pp. 211-241, 2001.
[http://dx.doi.org/10.1023/A:1013188200053]

[120] E.E. Alonso, and J-M. Pereira, "J. Vaunat and and S. Olivella, "A micro structurally based effective stress for unsaturated soils", *Geotechnique*, vol. 60, pp. 913-925, 2010.
[http://dx.doi.org/ 10.1680/geot.8.P.002]

[121] G. Della Vecchia, "C, Jommi and E. Romero, "A fully coupled elastic-plastic hydromechanical model for compacted soils accounting for clay activity", *Int. J. Numer. Anal. Methods Geomech.*, vol. 37, pp. 503-535, 2012.
[http://dx.doi.org/10.1002/nag.1116]

[122] D.A. Sun, D.C. Sheng, H.B. Cui, and S.W. Sloan, "A density-dependent elastoplastic hydro-mechanical model for unsaturated compacted soils", *Int. J. Numer. Anal. Methods Geomech.*, vol. 31, pp. 1257-1279, 2007.
[http://dx.doi.org/10.1002/nag.579]

[123] R.A. Rodrigues, and O.M. Volar, "Experimental study of the collapsible behavior of a tropical unsaturated soil", *Proceedings 5th Int. Conf. Unsat. Soils*, vol. 1, 2010pp. 353-357 Barcelona, Spain
[http://dx.doi.org/10.1201/b10526-47]

[124] A. Tarantino, and E. De Col, "Compaction behavior of clay", *Geotechnique*, vol. 58, pp. 199-213, 2008.
[http://dx.doi.org/10.1680/geot.2008.58.3.199]

[125] B. Caicedo, J. Tristancho, L. Thorel, and S. Leroueil, *Experimental and analytical framework for modeling soil compaction*. Engg. Geology, 2014.

[126] D.A. Sun, H. Matsuoka, and Y. Xu, "Collapse of compacted clay in suction-controlled triaxial cell", *Geotech. Test. J.*, vol. 27, pp. 362-370, 2004.

[127] J. H. F. Pereira, D. G. Fredlund, and M. P. Cardao Neto, *J. Geotech. Geoenviron. Eng.*, vol. 131, pp. 1264-1273, 2005.
[http://dx.doi.org/10.1061/(ASCE)1090-0241(2005)131:10(1264)]

[128] J.K. Mitchell, *Fundamentals of soil behavior*. John Wiley and sons, 1993.

[129] A. Anandarajah, and P.M. Amarasinghe, "Discrete-element study of the swelling behavior of Na-montmorillonite", *Geotechnique*, vol. 63, pp. 674-681, 2015.
[http://dx.doi.org/10.1680/geot.12.P.012]

[130] A. Gens, and E.E. Alonso, "A framework for the behavior of unsaturated expansive clays", *Can. Geotech. J.*, vol. 29, pp. 1013-1032, 1992.
[http://dx.doi.org/10.1139/t92-120]

[131] E.E. Alonso, J. Vaunat, and A. Gens, "Modelling the mechanical behavior of expansive clays", *Eng. Geol.*, vol. 54, pp. 173-183, 1999.
[http://dx.doi.org/10.1016/S0013-7952(99)00079-4]

[132] A. Lloret, V. Villar, M. Sánchez, A. Gens, X. Pintado, and E.E. Alonso, "Mechanical behavior of heavily compacted bentonite under high suction changes", *Geotechnique*, vol. 53, pp. 27-40, 2003.

[http://dx.doi.org/10.1680/geot.2003.53.1.27]

[133] E.E. Alonso, E. Romero, and C. Hoffmann, "Hydromechanical behaviour of compacted granular expansive mixtures: experimental and constitutive study", *Geotechnique,* vol. 61, pp. 329-344, 2011. [http://dx.doi.org/10.1680/geot.2011.61.4.329]

[134] W. Sun, D. Sun, and J. Li, "Elastoplastic modeling of hydraulic and mechanical behavior of unsaturated expansive soil", *Proc. GeoShangai 2010, Int. Conf.,* 2010pp. 119-127 [http://dx.doi.org/ 10.1061/41103(376)15]

[135] W. Sun, and D. Sun, "Coupled modeling of hydro-mechanical behavior of unsaturated compacted expensive soils", *Int. J. Numer. Anal. Methods Geomech.,* vol. 36, pp. 1002-1022, 2012. [http://dx.doi.org/ 10.1002/nag.1036]

[136] D. Mâsín, "Double structure hydromechanical coupling formalism and a model for unsaturated expansive clays", *Eng. Geol.,* vol. 165, pp. 73-88, 2013. [http://dx.doi.org/10.1016/j.enggeo.2013.05.026]

[137] J. Li, Z-H. Yin, Y. Cui, and Y. Hicher, "Work input analysis for soils with double porosity and application to the hydro-mechanical modeling of unsaturated clays", *Can. Geotech. J.,* 2016. [http://dx.doi.org/10.1139/cgj-2015-0574]

[138] G.E. Blight, The time-rate of have of structures on expansive clays.*Moisture equilibria and moisture changes in soils beneath covered areas.* Butterworths: Sydney, 1965, pp. 78-87.

[139] G. Kassiff, and A. Ben Shalom, "Experimental relationship between swell pressure and suction", *Geotechnique,* vol. 2, pp. 1245-1255, 1971.

[140] D.G. Fredlund, J.U. Hasam, and H. Filson, "The prediction of total heave", *Proc. Fourth Int. Conf. Exp. Soils,* 1980pp. 1-17 Denver, Colorado

[141] H.Q. Vu, and D.G. Fredlund, "Numerical modelling of two dimensional heave for slabs-on-ground and shallow foundations", *Proc. 56th Can. Geotech. Conf.,* 2003pp. 220-227 Winnipeg

[142] H. Tu, and S.K. Vanapalli, "Prediction of the variation of swelling pressure and one dimensional heave of expansive soils with respect to suction using the soil-water retention curve as a tool", *Can. Geotech. J.,* 2016. [http://dx.doi.org/10.1139/cgj-2015-0222]

[143] R. Monroy, L. Zdravkovic, and A. Ridley, "Evolution of microstructure in compacted London Clay during wetting and loading", *Geotechnique,* vol. 60, pp. 105-119, 2010. [http://dx.doi.org/10.1680/ geot.8.P.125]

[144] C. Hoffmann, E.E. Alonso, and E. Romero, "Hydro-mechanical behaviour of bentonite pellet mixtures", *Phys. Chem. Earth,* vol. 32, pp. 832-849, 2007. [http://dx.doi.org/10.1016/j.pce.2006.04.037]

[145] M.V. Villar, and A. Lloret, "Influence of dry density and water content on the swelling of a compacted bentonite", *Appl. Clay Sci.,* vol. 39, pp. 38-49, 2008. [http://dx.doi.org/10.1016/j.clay.2007.04.007]

[146] H. Nowamooz, and F. Masrouri, "Mechanical Behavior of expansive soils after several drying-wetting cycles", *Geomech. Geoengng. Int. J.,* vol. 5, pp. 213-221, 2010. [http://dx.doi.org/10.1080/17486025. 2010.521588]

[147] E. Romero, G. Della Vecchia, and C. Jommi, "An insight into the water retention properties of compacted clayey soils", *Geotechnique,* vol. 61, pp. 313-328, 2011. [http://dx.doi.org/10.1680/geot. 2011.61.4.313]

[148] H. Arroyo, E. Rojas, M.L. Pérez-Rea, J. Horta, and J. Arroyo, "A porous model to simulate the evolution of the soil-water characteristic curve with volumetric strains", *C. R. Mec.,* 2015. [http://dx.doi.org/10.1016/j.crme.2015.02.001]

[149] S.A. Habib, "Lateral pressure of unsaturated expansive clay in looped stress path", *Proc. First Int.*

Conf. Unsat. Soils, 1995pp. 95-100

[150] E.E. Alonso, A. Lloret, A. Gens, and D.Q. Yang, "Experimental behavior of highly expansive double-structure clay", *First Int. Conf. Unsat. Soils,* vol. 1, 1995pp. 11-16 Paris, France

[151] N. Lu, J.W. Godt, and D.T. Wu, "A closed-form equation for effective stress in unsaturated soil", *Water Resour. Res.,* vol. 46, 2010.
[http://dx.doi.org/10.1029/2009WR008646]

[152] E. Rojas, M.L. Pérez-Rea, T. López-Lara, J.B. Hernández, and J. Horta, "Use of effective stresses to model the collapse upon wetting of unsaturated soils", *J. Geotech. Geoenviron. Eng.,* vol. 141, pp. 1-13, 2015.
[http://dx.doi.org/10.1061/(ASCE)GT.1943-5606.0001251]

[153] S.J. Wheeler, "Inclusion of specific water volume within an elastoplastic model for unsaturated soil", *Can. Geotech. J.,* vol. 33, pp. 35-53, 1996.
[http://dx.doi.org/10.1139/t96-023]

[154] C. Jommi, "Remarks on the constitutive modelling of unsaturated soils. Experimental evidence and theoretical approaches in unsaturated soils", *Proc. of the Int. Workshop on Unsat. Soils,* 2000pp. 139-153 Rotterdam

[155] M.S.R. Buisson, and S.J. Wheeler, "Inclusion of hydraulic hysteresis in a new elastoplastic framework for unsaturated soils", *Proc. of the Int. Workshop Unsat. Soils,* 2000pp. 109-119 Rotterdam

[156] D.A. Sun, D. Sheng, and S.W. Sloan, "Elastoplastic modeling of hydraulic and stress-strain behavior of unsaturated soil", *Mech. Mater.,* vol. 39, pp. 212-221, 2007.
[http://dx.doi.org/10.1016/j.mechmat. 2006.05.002]

[157] K. Kawai, S. Kato, and D. Karube, "The model of water retention curve considering effects of void ratio", *Proc. of the Asian Conf. Unsat. Soils,* 2000pp. 329-334 Singapore

[158] D. Gallipoli, S.J. Wheeler, and M. Karstunen, "Modelling the variation of degree of saturation in a deformable unsaturated soil", *Geotechnique,* vol. 53, pp. 105-112, 2003.
[http://dx.doi.org/10.1680/ geot.2003.53.1.105]

[159] M. Nuth, and L. Laloui, "Advances in modeling hysteretic water retention curve in deformable soils", *Comput. Geotech.,* vol. 35, pp. 835-844, 2008.
[http://dx.doi.org/10.1016/j.compgeo.2008.08.001]

[160] A. Tarantino, "A water retention model for deformable soils", *Geotechnique,* vol. 59, pp. 751-762, 2009.
[http://dx.doi.org/10.1680/geot.7.00118]

[161] D. Masín, "Predicting the dependency of a degree of saturation on void ratio and suction using effective stress principle for unsaturated soils", *Int. J. Numer. Anal. Methods Geomech.,* vol. 34, pp. 73-90, 2010.

[162] D. Sheng, and A.N. Zhou, "Coupling hydraulic with mechanical models for unsaturated soils", *Can. Geotech. J.,* vol. 48, pp. 826-840, 2011.
[http://dx.doi.org/10.1139/t10-109]

[163] D. Gallipoli, "A hysteretic soil-water retention model accounting for cyclic variations of suction and void ratio", *Geotechnique,* vol. 62, pp. 605-616, 2012.
[http://dx.doi.org/10.1680/geot.11.P.007]

[164] S. Salager, M. Nuth, A. Ferrari, and L. Laloui, "Investigation into water retention behavior of deformable soil", *Can. Geotech. J.,* vol. 50, pp. 200-208, 2013.
[http://dx.doi.org/10.1139/cgj-2011-0409]

[165] C. Zhou, and C.W.W. Ng, "A new and simple stress-dependent water retention model for unsaturated soils", *Comput. Geotech.,* vol. 62, pp. 216-222, 2014.
[http://dx.doi.org/10.1016/j.compgeo.2014.07.012]

[166] N. Khalili, M.A. Habte, and S. Zargarbashi, "A fully coupled flow deformation model for cyclic analysis of unsaturated soils including hydraulic and mechanical hysteresis", *Comput. Geotech.*, vol. 35, pp. 872-889, 2008.
[http://dx.doi.org/10.1016/j.compgeo.2008.08.003]

[167] R. Brooks, and A. Corey, Hydraulic properties of porous media.*Hydrology Paper No. 3*. Colorado State University: Fort Collins, CO, 1964.

[168] E.E. Alonso, N.M. Pinyol, and A. Gens, "Compacted soil behavior, initial state, structure and constitutive modeling", *Geotechnique,* vol. 63, pp. 463-478, 2013.
[http://dx.doi.org/10.1680/geot.11.P.134]

[169] S. Jayanth, K. Iyer, and D.N. Singh, "Influence of drying- and wetting- cycles on SWCC of fine-grained soils", *J. Test. Eval.,* vol. 40, pp. 376-386, 2012.
[http://dx.doi.org/10.1520/JTE104184]

[170] S. Jayanth, K. Iyer, and D.N. Singh, "Continuous determination of drying-path SWRC of fine-grained soils", *Geomech. Geoengng. Int. J.,* vol. 8, pp. 28-35, 2013.
[http://dx.doi.org/10.1080/17486025.2012. 727034]

[171] C.F. Chiu, and C.W.W. Ng, "Coupled water retention and shrinkage properties of a compacted silt under isotropic and deviatoric stress paths", *Can. Geotech. J.,* vol. 49, pp. 928-938, 2012.
[http://dx.doi.org/ 10.1139/t2012-055]

[172] C.W.W. Ng, and Y.W. Pang, "Experimental investigations of the soil-water characteristics of a volcanic soil", *Can. Geotech. J.,* vol. 37, pp. 1252-1264, 2000.
[http://dx.doi.org/10.1139/t00-056]

[173] E. Rojas, and O. Chavez, "Volumetric behavior of unsaturated soils", *Can. Geotech. J.,* vol. 50, pp. 209-222, 2013.
[http://dx.doi.org/10.1139/cgj-2012-0341]

[174] R. Hu, Y-F. Chen, H-H. Liu, and C-B. Zhou, "A coupled two-phase fluid flow and elastoplastic deformation model for unsaturated soils: theory, implementation, and application", *Int. J. Numer. Anal. Methods Geomech.,* vol. 40, pp. 1023-1058, 2002.
[http://dx.doi.org/10.1002/nag.2473]

[175] E. Liu, H-S. Yu, G. Deng, J. Zhang, and S. He, "Numerical analysis of seepage-deformation in unsaturated soils", *Acta Geotech.,* vol. 9, pp. 1045-1058, 2014.
[http://dx.doi.org/10.1007/ s11440-014-0343-y]

[176] J. Choo, J.A. White, and R.I. Borja, "Hydromechanical modeling of unsaturated flow in double porosity media", *Int. J. Geomech.,* 2016.
[http://dx.doi.org/10.1061/(ASCE)GM.1943-5622.0000558]

[177] M. Lloret-Cabot, S.J. Wheeler, and M. Sánchez, "A unified mechanical and retention model for saturated and unsaturated soil behavior", *Acta Geotech.,* vol. 12, pp. 1-21, 2017.
[http://dx.doi.org/ 10.1007/s11440-016-0497-x]

[178] X. Song, K. Wang, and M. Ye, "Localized failure in unsaturated soils under non-isothermal conditions", *Acta Geotech.,* vol. 13, pp. 73-85, 2018.
[http://dx.doi.org/10.1007/s11440-017-0534-4]

[179] B. Loret, and N. Khalili, "Effective stress elastic-plastic model for unsaturated porous media", *Mech. Mater.,* vol. 34, pp. 97-116, 2002.
[http://dx.doi.org/10.1016/S0167-6636(01)00092-8]

[180] D. Sun, D. Sheng, and S. Sloan, "Elastoplastic modelling of hydraulic and stress-strain behavior of unsaturated soils", *Mech. Mater.,* vol. 39, pp. 212-221, 2007.
[http://dx.doi.org/10.1016/j.mechmat. 2006.05.002]

[181] D. Sun, W. Sun, and L. Xiang, "Effect of degree of saturation on mechanical behavior of unsaturated

soils and its elastoplastic simulation", *Comput. Geotech.,* vol. 37, pp. 678-688, 2010.
[http://dx.doi.org/ 10.1016/j.compgeo.2010.04.006]

[182] W.T. Solowski, and S.W. Sloan, "Equivalent stress approach in creation of elastoplastic constitutive models for unsaturated soils", *Int. J. Geomech.,* 2013.
[http://dx.doi.org/ 10.1061/(ASCE)GM.1943-5622.0000368]

[183] R. Hu, Y-H. Chen, H-H. Liu, and C-B. Zhou, "A coupled stress-strain and hydraulic hysteresis model for unsaturated soils: Thermodynamics analysis and model evaluation", *Comput. Geotech.,* vol. 63, pp. 159-170, 2015.
[http://dx.doi.org/10.1016/j.compgeo.2014.09.006]

[184] A. Zhou, and D. Sheng, "An advanced hydro-mechanical constitutive model for unsaturated soils with different initial densities", *Comput. Geotech.,* vol. 63, pp. 46-66, 2015.
[http://dx.doi.org/ 10.1016/j.compgeo.2014.07.017]

[185] T. Ma, C. Wei, H. Wei, and W. Li, "Hydraulic and mechanical behavior of unsaturated silt: experimental and theoretical characterization", *Int. J. Geomech.,* vol. 16, pp. 1-13, 2015.

[186] D. Masín, and N. Khalili, "A hypoplastic model for mechanical response of unsaturated soils", *Int. J. Numer. Anal. Methods Geomech.,* vol. 32, pp. 1903-1926, 2008.
[http://dx.doi.org/10.1002/nag.714]

[187] W. Fuentes, and Th. Triantafyllidis, "Hydro-mechanical hypoplastic models for unsaturated soils under isotropic stress conditions", *Comput. Geotech.,* vol. 51, pp. 72-82, 2013.
[http://dx.doi.org/ 10.1016/j.compgeo.2013.02.002]

[188] D. Manzanal, M. Pastor, and J. A. Fernández Merodo, *Generalized plasticity state parameter-based model for saturated and unsaturated soils. Part II: Unsaturated soil modeling,* .

[189] G.M. Rostisciani, G. Sciarra, F. Casini, and A. Desideri, "Hydro-mechanical response of collapsible soils under different infiltration events", *Int. J. Numer. Anal. Methods Geomech.,* vol. 39, pp. 1212-1234, 2015.
[http://dx.doi.org/10.1002/nag.2359]

[190] R. Hu, H-H. Liu, Y. Chen, C. Zhou, and D. Gallipoli, "A constitutive model for unsaturated soils with consideration of inter-particle bonding", *Comput. Geotech.,* vol. 59, pp. 127-144, 2014.
[http://dx.doi.org/ 10.1016/j.compgeo.2014.03.007]

[191] M.D. Fredlund, G.W. Wilson, and D.G. Fredlund, "Use of the grain size distribution for estimation of the soil-water characteristic curve", *Can. Geotech. J.,* vol. 39, pp. 1103-1117, 2002.
[http://dx.doi.org/ 10.1139/t02-049]

[192] A. Russell, and N. Khalili, "A unified bounding surface model for unsaturated soils", *Int. J. Numer. Anal. Methods Geomech.,* vol. 30, pp. 181-212, 2006.
[http://dx.doi.org/10.1002/nag.475]

[193] M. Feng, and D.G. Fredlund, "Hysteresis influence associated with thermal conductivity sensor measurements",

[194] Q. Wang, D.E. Pufahl, and D.G. Fredlund, "A study of critical state on an unsaturated silty soil", *Can. Geotech. J.,* vol. 39, pp. 213-218, 2002.
[http://dx.doi.org/10.1139/t01-086]

[195] T.M. Thu, H. Rahardjo, and E-C. Leong, "Soil-water characteristic curve and consolidation behavior for a compacted silt", *Can. Geotech. J.,* vol. 44, pp. 266-275, 2007.
[http://dx.doi.org/10.1139/t06-114]

[196] M. Nuth, and L. Laloui, "Effective stress concept in unsaturated soil: Clarification and validation of a unified framework", *Int. J. Numer. Anal. Methods Geomech.,* vol. 32, pp. 771-801, 2008.
[http://dx.doi.org/ 10.1002/nag.645]

[197] N. Lu, Y.K. Kim, S.J. Lee, and S.R. Lee, "Relationship between the soil-water characteristic curve and

suction stress characteristic curve: experimental evidence from residual soils", *J. Geotech. Geoenviron. Eng.,* vol. 138, pp. 47-57, 2012.
[http://dx.doi.org/10.1061/(ASCE)GT.1943-5606.0000564]

[198] Y.F. Dafalias, and L.R. Hermann, "Bounding surface plasticity I: Mathematical foundation and hypoplasticity", *J. Eng. Mech.,* vol. 112, pp. 966-987, 1986.
[http://dx.doi.org/10.1061/ (ASCE)0733-9399(1986)112:9(966)]

[199] K. Hashiguchi, "Subloading surface model in unconventional plasticity", *Int. J. Solids Struct.,* vol. 25, pp. 917-945, 1989.
[http://dx.doi.org/10.1016/0020-7683(89)90038-3]

[200] K. Been, and M.G. Jefferies, "A state parameter for sands", *Geotechnique,* vol. 35, pp. 99-11, 1985.
[http://dx.doi.org/10.1680/geot.1985.35.2.99]

[201] S. Jockovic, and M. Vukicevic, "Bounding surface model for overconsolidated clays with new state parameter formulation of hardening rule", *Comput. Geotech.,* vol. 83, pp. 16-29, 2017.
[http://dx.doi.org/10.1016/j.compgeo.2016.10.013]

[202] M.M. Futai, and S.S. Almeida, "Yield, strength, and critical state behavior of a tropical saturated soil", *J. Geotech. Geoenviron. Eng.,* vol. 130, pp. 1169-1179, 2004.
[http://dx.doi.org/10.1061/ (ASCE)1090-0241(2004)130:11(1169)]

[203] Y.J. Cui, and P. Delage, "Yielding and plastic behavior of an unsaturated compacted silt", *Geotechnique,* vol. 46, pp. 291-311, 1996.
[http://dx.doi.org/10.1680/geot.1996.46.2.291]

[204] A.A. Garakani, S.M. Haeri, A. Khosravi, and G. Habibagahi, "Hydro-mechanical behavior of undisturbed collapsible loessial soils under different stress state conditions", *Eng. Geol.,* vol. 195, pp. 28-41, 2015.
[http://dx.doi.org/10.1016/j.enggeo.2015.05.026]

BIBLIOGRAPHY

This book has been prepared using material from the following publications

E. Rojas, and F Rojas, "Modeling hysteresis of the soil-water characteristic curve", *Soil Found.*, vol. 45, pp. 135-146, 2005. [With permission from Elsevier.]
[http://dx.doi.org/10.3208/sandf.45.3_135]

E.E. Alonso, E. Rojas, and N.M Pinyol, "Unsaturated soil mechanics", Reunión Nacional de Mecánica de Suelos, Aguascalientes, Mexico, Especial Volume, pp. 117-205, 2008. With permission from Sociedad Mexicana de Ingeniería Geotécnica

E Rojas, "Equivalent stress equation for unsaturated soils, I: Equivalent stress", *Int. J. Geomech.*, vol. 8, pp. 285-290, 2008. [With permission from ASCE.]
[http://dx.doi.org/10.1061/(ASCE)1532-3641(2008)8:5(285).]

E Rojas, "Equivalent stress equation for unsaturated soils, II: Solid-porous model", *Int. J. Geomech.*, vol. 8, pp. 291-299, 2008. [With permission from ASCE.]
[http://dx.doi.org/10.1061/(ASCE)1532-3641(2008)8:5(291).]

E. Rojas, J. Horta, T. López-Lara, and J.B Hernández, *A probabilistic solid-porous model to determine the shear strength of unsaturated soils*, 2011.

E. Rojas, M.L. Pérez-Rea, G. Gallegos, and J Leal, "A porous model for the interpretation of mercury intrusion porosimetry tests", *J. Porous Media*, vol. 15, pp. 517-530, 2012. [With permission from John Wiley and Sons Ltd.]
[http://dx.doi.org/10.1615/JPorMedia.v15.i6.20]

J. Horta, E. Rojas, M.L. Pérez, T. López-Lara, and J.B Hernández, "A random porous model to simulate the retention curve of soils", *Int. J. Numer. Anal. Methods Geomech.*, vol. 37, pp. 932-944, 2012. [With permission from John Wiley and Sons.]
[http://dx.doi.org/10.1002/nag.1133.]

E. Rojas, J. Horta, T. López Lara, and J. B Hernández, *A solid-porous model to simulate the retention curves of soils*, .

E. Rojas, and O Chávez, "Volumetric behavior of unsaturated soils", *Canadian Journal of Geomechanics*, vol. 50, pp. 209-222, 2013. [With permission from NRC Research Press.]

E. Rojas, T. Lopez-Lara, J.B. Hernández, and J Horta, "Use of effective stresses to model the collapse upon wetting in unsaturated soils", *J. Geotech. Geoenviron. Eng.*, vol. 141, pp. 1-13, 2015. [With permission from ASCE.]
[http://dx.doi.org/10.1061/(ASCE)GT.1943-5606.0001251]

E. Rojas, O. Chávez, H. Arroyo, T. López-Lara, J.B. Hernández, and J Horta, "Modeling the dependency of the soil-water retention curve on the volumetric deformation. ", *Int. J. Geomech.*, vol. 17, pp. 1-13, 2017. [With permission from ASCE.]
[http://dx.doi.org/10.1061/(ASCE)GM.1943-5622.0000678.]

E. Rojas, O. Chávez, and H Arroyo, "Modeling the behavior of expansive soils using effective stresses", *Int. J. Geomech.*, vol. 17, pp. 1-15, 2017. [With permission from ASCE.]
[http://dx.doi.org/10.1061/(ASCE)GM.1943-5622.0000943.]

LIST OF ABBREVIATIONS

BBM	Barcelona Basic Model
CSL	critical state line
DEM	distinct element method
Ex	Experimental
GSD	grain size distribution
LCYS	loading collapse yield surface
MIP	mercury intrusion porosimetry
MCCM	modified Cam-Clay model
N	numerical
NL	neutral line
PSD	pore size distribution
SDYS	suction decrease yield surface
SEM	scanning electron micrographs
SIYS	suction increase yield surface
SMIG	Sociedad Mexicana de Ingeniería Geotécnica
SWRC	soil water retention curve
URL	unloading-reloading line
WP	wetting path

GLOSSARY

A total area of a cross section of an unsaturated soil

A_a total area of the cross section where air reacts

A_w total area of the cross section where water reacts

A_s total area of the cross section where solids react

A^s area of the saturated fraction

A^u area of the unsaturated fraction

A_a^u area of the unsaturated fraction where air reacts

$A_{\bar{s}}^s$ area of solids of the saturated fraction

$A_{\bar{s}}^u$ area of solids of the unsaturated fraction

$A_{\bar{s}a}^u$ horizontal projection of the peripheral area of solids in contact with air in the unsaturated fraction

$A_{\bar{s}w}^{\bar{s}}$ horizontal projection of the peripheral area of solids subject to water pressure in the saturated fraction

$A_{\bar{s}w}^u$ horizontal projection of the peripheral area of solids in contact with water in the unsaturated fraction

A_v^u area of voids in the unsaturated fraction

A_w^s area of the saturated fraction where water reacts

A_w^u area of the unsaturated fraction where water reacts

a contact area between solid particles per unit area of material

$B(R_C)$ probability function of the relative volume of bonds

$B_A(R_C)$ adjusted probability function for bonds

C connectivity (number of bonds converging at one site)

C_c cluster coefficient

C_e compressibility of the solid structure of the soil

$C_{\bar{s}}$ compressibility of the solid material comprising the particles

C_{SB} fraction of sites and bonds in closed clusters

c soil cohesion

D diameter of a spherical particle

D_e equivalent diameter of clay particles

D_r relative density

d diameter of a pore

e void ratio

e_{max} maximum void ratio of expanded soil

e_{min} minimum void ratio of severely compacted soils

f^d dry soil fraction

f^s saturated soil fraction

f^u unsaturated soil fraction

F_{BD}^S saturation factor for bonds during drying

F_{BI}^S saturation factor for bonds during wetting

F_{BID}^S saturated wetting factor for bonds after a drying inversion

F_{SD}^S saturated draying factor

F_{SDI}^S saturated wetting factor for sites after a drying inversion

F_{SD}^d dry factor for cavities in the solid unit during a drying process

F_{SI}^s saturation factor for cavities during wetting

F_{SID}^s drying saturation factor for cavities after drying inversion

G_{BDR} bonds invaded by gas at the end of a drying inversion

G_{BD1} probability of an external bond to be filled with gas during drying

G_{BI} probability of the bond to be filled with gas

G_{PD} probability of a solid to be surrounded exclusively by pores filled with gas during drying

G_{PI} probability of a solid to be surrounded exclusively by pores filled with gas during wetting

G_{SDR} sites invaded by gas at the end of a drying inversion

G_{SI} probability of a cavity to be filled with gas

k parameter related to the effective stress for saturated soils

L_{BD} probability of a bond to be liquid-filled at drying

L_{BDR} probability for a bond to be liquid-filled at the moment of a drying inversion

L_{BI} probability for a bond to be liquid-filled after an inversion in drying

L_{BD1} probability of an external bond to be liquid-filled

L_{BI} probability for a bond to saturate during a wetting process

L_{BID} probability for a bond to be liquid-filled at drying after a wetting inversion

L_{BIR} probability for a bond to be liquid-filled at the moment of a wetting inversion

L_{BI1} probability of an external bond to be saturated

L_{SD} probability of a cavity to saturated during a drying process

L_{SDI} probability for a cavity to be liquid-filled after an inversion in drying

L_{SDR} probability for a site to be liquid-filled at the moment an inversion in drying

L_{SD1} probability of surrounding sites to be liquid filled during a drying process

L_{SI} probability for a cavity to saturate during a wetting process

L_{SID} probability for a site to be liquid-filled at drying after a wetting inversion

L_{SIR} probability for a site to be liquid-filled at the moment of a wetting inversion

L_{sn} probability of a surrounding site to be liquid filled

$L_{\tilde{S}D}$ probability that all pores surrounding a solid are saturated during a drying process

$L_{\tilde{S}I}$ probability that all pores surrounding a solid are saturated during a wetting process

N_{rMi} real number of macropores of size i

N_{TMi} total number of macropores of size i

n soil porosity

$\bar{p}$ mean net stress

p' mean effective stress

p mean total stress

P_{atm} atmospheric pressure

q deviator stress

$\bar{R}$ mean size

$\bar{R}_B$ mean size for bonds

R_c critical radius or maximum size of a pore to be filled with water at certain suction

$\bar{R}_S$ mean size for sites

R_{vM} Relative volume factor of macropores

R_{vM0} initial relative volume factor of macropores

R_{vM1} current relative volume of macropores

R_{vB0} initial relative volume factor of macrobonds

R_{vB1} current relative volume of macrobonds

r_B radius of a bond

$r_{\tilde{S}}^{s}$ ratio of the volume of saturated solids to the total volume of solids

r_{v}^{s} ratio of the volume of saturated voids to the total volume of voids

r_s radius of a site

$S(R_C)$ probability function of the relative volume of cavities

$S_A(R_C)$ adjusted probability function for sites

S_f shape factor for the solid particles

S_w degree of saturation

$S_{wD}(R_C)$ degree of saturation at drying

$S_{wI}(R_C)$ degree of saturation at wetting

S_{w}^{u} degree of saturation of the unsaturated fraction

s soil suction

s_m suction in a pore filled with mercury

s_w suction in a pore filled with water

T_s liquid-gas interfacial tension

T_{sm} superficial tension for mercury

T_{sw} superficial tension for water

u pore water pressure for saturated soils

u_a air pressure

u_w water pressure

V total volume of a portion of soil

V_a air volume

V_B volume of bonds

V_{BI}^S volume of saturated bonds during wetting

V_{BID}^S volume of saturated bonds during drying after a wetting inversion

V_{BR} volume of all bonds of size R

V_B^d volume of bonds of the dry fraction

V_B^s volume of saturated bonds

$V_{\bar{s}}$ volume of solids

$V_{\bar{s}}^d$ volume of solids of the dry fraction

V_s volume of cavities

V_S^d volume of cavities of the dry fraction

V_S^s volume of saturated cavities

V_{SI}^s volume of saturated cavities during a wetting process

V_{SID}^S volume of saturated sites during drying after a wetting inversion

$V_{SR}(R)$ volume of all sites of size R

V_B^d volume of dry bonds

$V_{BD}^s(R_C)$ volume of saturated bonds during drying

$V_{BDI}^s(R_C)$ volume of saturated bonds at wetting after a drying inversion

$V_{BI}^s(R_C)$ volume of saturated bonds at wetting

$V_{BR}(R)$ volume of bonds of size R

$V_{\bar{s}I}^s$ volume of saturated solids

$V_{RB}(R)$ relative volume of bonds of size R

$V_{RS}(R)$ relative volume of cavities of size R

$V_{SR}(R)$ volume of all sites of size R

$V_{SD}^s(R_C)$ volume of saturated sites at drying

$V_{SDI}^s(R_C)$ volume of saturated sites at the end of a drying inversion

$V_{SI}^s(R_C)$ volume of saturated cavities during wetting

V_v volume of voids of the soil sample

V_w water volume

V^s volume of the saturated fraction of the sample

V^u volume of the unsaturated fraction of the sample

V_a^u volume of air of the unsaturated fraction

V_s^s volume of solids of the saturated fraction

V_s^u volume of solids of the unsaturated fraction

V_{sa}^u volume of solids influenced by the pressure of air in the unsaturated fraction

V_{sw}^u volume of solids influenced by the pressure of water in the unsaturated fraction

V_v^s volume of voids of the saturated fraction

V_v^u volume of voids of the unsaturated fraction

V_w^s volume of water of the saturated fraction

V_w^u water volume of the unsaturated fraction

v specific volume

χ Bishop's effective stress parameter

ΔV_v^p irrecoverable reduction of the volume of voids

δ_{ij} Kronecker's delta ($\delta_{ij} = 1$, $i = j$; $\delta_{ij} = 0$ $i \neq j$)

θ_m contact angles of mercury with the minerals of soil

θ_w contact angles of water with the minerals of soil

κ fitting parameter

κ_e unloading-reloading compression index

λ_e virgin loading compression index

λ_{cex} collapsing-expansion index

λ_{ex} expansion index

σ total stress

δ standard deviation

δ_B standard deviation for bonds

δ_S standard deviation for sites

σ' effective stress

σ_s^* matric suction stress

σ_i principal total stress in direction i

σ_i' principal effective stress in direction i

σ_{ij} total stress tensor

σ_{ij}' effective stress tensor

$\overline{\sigma}$ net stress

τ shear stress

φ internal friction angle

ψ friction angle of the material comprising the solid particles

SUBJECT INDEX

*9 7 8 1 6 8 1 0 8 7 0 0 9 *